Civil War Journal of a Union Soldier

From the journal of Harmon Camburn (1861-1864)

PRESENTED BY P.C. ZICK

ISBN: 0988878232
ISBN-13: 978-09888782-3-5

IN MEMORIUM

To all the men and women who sacrificed home, families, and life for both sides during the Civil War. May you all rest in peace.

CONTENTS

ACKNOWLEDGMENTS

While the journal of Harmon Camburn was intended as an account to share with his children, today it represents a chapter in the history of the United States. My great grandfather's descriptive story of his military duty is both beautiful and horrific. I thank him for leaving this rich legacy that represents a very personal view of the cost of war.

~P.C. (Patricia Camburn) Zick.

INTRODUCTION

Less than two weeks after the first shot rang out at Fort Sumter on April 12, 1861, my great grandfather, Harmon Camburn, entered into the service of the Union army. His service for the Union cause lasted three years. It ended with a bullet through his lung, but against great odds, he survived. Not only did he survive, but he also left this journal.

Many years after the Civil War ended, his children—one of whom was my grandfather, Arthur Thomas Camburn—asked him to tell the story of his years as a Union soldier in the 2nd Michigan Infantry. Those written memories existed in their original handwritten form until 1975 when my cousin, Helen Ann Boughton Noble, typed them and made copies for all of his progeny. For years, I've kept this "book" on my bookshelves, reading the pages occasionally and wishing they could receive a larger audience. One day, I decided the time had arrived to publish the journal. This personal and fascinating piece of history shows us a close view of life for a soldier from the mundane to the profane and from the petty to the fatal.

Harmon Camburn proudly served his country, and through his journal, expressed the conundrum that existed as soldiers fought against their fellow countrymen. Unfortunately, I do not have any photos of my great grandfather.

A note on editing of the journal: I did as little as possible to keep the style of the time. The word order and usage may seem formal and out of date, but it's also historical. I tried not to change too much. If I edited my great-grandfather's journal at all, it was to help the reader understand the meaning, but that only depends on my understanding of the typed version. Several times words made no sense in sentences but that could be the result of typos when by cousin when she typed it originally. Since she is deceased, I can't ask her about her edits and interpretations.

The journal is very specific to my great-grandfather's experience during his three-year stint as a Union soldier. He doesn't give an account of the bigger picture of the war most of the time. I believe there are several reasons for this lack of information. First, the soldiers probably didn't know what was happening outside of their narrow world. According to my great-grandfather, at times they didn't even know where they were marching or for what reason. Second, he wrote the journal within a few decades of the war, before the Great World Wars of the Twentieth Century. He most likely assumed his audience had an intimate knowledge of the details of the Big War.

Now more than 150 years later, I realized I'd forgotten much of what I'd learned about the Civil War and its battles between the northern and southern armies. I began researching and found it fascinating to connect all the parts between the soldiers and the war as portrayed in the movies and history books. He suggests several times that the outside world of newspapers, politicians, and civilians had no idea of what was happening on the ground. It must have been demoralizing to be in the rain and cold with bullets flying nearby, to realize that supporters were spending a lot of time criticizing the military machine. All the while, those fighting men were often starving and dying.

I've inserted short blurbs about the greater war to correspond with the dates from the journal. With some of the more interesting characters, I sometimes added tidbits of information. My insertions are in bold type encased with brackets. There were footnotes in the typewritten manuscript, but I don't know if my cousin added those or if Harmon Camburn added them. I've included them in parenthesis after the appropriate journal text. I added the full names of generals, but not of lower ranks. Each of the chapter titles includes the date at the beginning of the narrative. It provides a marker for time.

I found obituaries for Harmon Camburn in a file that came from one of my aunts. They were clipped out of newspapers, so I don't have the full name of the newspapers. They differ somewhat in the information provided. I included them at the end of the journal to give an idea of the life of my great-grandfather after the Civil War. I didn't edit these but typed them exactly as they appeared in the paper. It's interesting to see how language and mechanics have changed in the century since these accounts were printed. Anyone who remembers him is also deceased, as are the folks who knew him from stories passed down. It's been said that when a person dies, a library burns down. I'm grateful my great grandfather left this one record. Other than this journal and the obituaries, I know very little about him. I know even less about his wife—my great grandmother—who he married the year he came home from the war. There is no record of how they met, and my great-grandfather doesn't refer to anyone special in his life during the course of the war.

With great pleasure, I share the library of Harmon Camburn.

Patricia Camburn Zick
Beaver County, Pennsylvania
October 2013

EXCUSE

At the earnest solicitation of my children for stories of my past life, I promised to write in my leisure moments some incidents for their perusal, as I could not always respond at their desire with an interesting story.

Memory, aided by a brief diary that I kept during my service in the army, are all the sources I have to draw from.

While I shall endeavor to be strictly truthful, I must of necessity speak of things from a personal standpoint, or as I saw them. And the impressions will be those made upon me by passing events.

If what I write meets the eye of others than those for whom they are intended, I have only this to say: It was only written for my children. And if I confer upon them as much pleasure as I shall take in gratifying them, I shall feel amply repaid.

Harmon and Eliza Camburn

CHAPTER 1 – 1842-1861

I was born on a farm in the township of Franklin, County of Lenawee, State of Michigan, February 4, 1842.

I successfully passed through the measles, mumps, and other diseases of children until I was six years and seven months old when my mother died leaving me to the care of older sisters, and afterward to a stepmother when my father remarried.

Being the youngest of fourteen children, I was subjected to the usual process of teasing and tantalization until my disposition was as near the reverse of angelic as was at all desirable. **[Family tree and obituary list sixteen children. One child, a twin of firstborn, died in infancy. Another sibling, Ann Mariah Camburn died when he was eight, according to the family tree.]** Those whose special delight it was to see me angry had need to be prepared to avoid clubs or stones that were hurled at them, and some yet bear scars that are the results of these outbreaks.

Prophecies of hanging for manslaughter were very familiar to me; and the horrors of prison life were often painted for my benefit.

The Mother's watchful care was needed to curb the wayward disposition of the child, and also to repress the cruel irritators who thought it fun to witness the mad freaks of a passionate child.

I continued to live on the farm where I was born, working summers and attending district school winters until I was seventeen years old. Having advanced as far as I could in my studies in the district school, my Father sent me to the Raisin Institute in the winter of 1858 and '59.

I returned to the farm for the summer, and being tall and strong, I swung the grain cradle and hay scythe most of the season.

My older brother took the farm to work, and I being of no use to my father at home, it was arranged that I should continue my studies at Raisin Institute in the fall, and thereafter depend on my own resources.

Feeling that I was large enough and competent in an educational point of view, I engaged to teach a district school during the winter.

After conducting my school to a successful close, I returned to the Institute for the spring term. At the close of school, I hired out to work on a farm during the summer vacation and was again at the Institute for the fall term. In the winter, I taught another term of district school, which was successful.

It had been my intention to continue working summers and teaching winters, and with the money so earned to work my way through college. But the political ferment that had been so long brewing between the North and South began to assume proportions that boded trouble to the nation. The threat of the Southerners to dissolve the Union was being discussed all over the country.

Rumors of troops being raised to resist the government began to reach us.

The excitement was growing so intense that little else was talked of in the family circle, on the streets, or in public gatherings. Resistance to southern outrages was even preached from the pulpit.

While watching the course of events with absorbing interest, I had made up my mind to embrace the first opportunity, should there be any call, to enlist to help put down the coming rebellion, which no one thought would be more than a summer campaign.

Notwithstanding the boasts and threats of the South, the firing on Fort Sumter fell like a thunderbolt on the people.

[On April 12, 1861, Confederate warships fired on Union soldiers at Fort Sumter, South Carolina, to begin the Civil War.]

Immediately, the whole North began to organize military companies; and war meetings were held everywhere.

Then came the call of President Lincoln for seventy-five thousand men for three months. Michigan was asked for one regiment of ten companies under this call. Two companies were started in Adrian: the Hardee Cadets and the Adrian Guards.

The Guards being the oldest company, I thought they would be the first accepted; and consequently chose that company, thinking that perhaps that would be the last chance I would ever have to serve my country as a soldier.

On the morning of April 20, 1861, my father said to me at breakfast, "If you will harness the horse, we will go to Adrian and hear the latest news from Washington."

On our arrival at Adrian, Father left me at liberty while he transacted some necessary business, and I made my way directly to the recruiting office of the Adrian Guards where I signed a pledge to enlist in the company for three months. I soon met Father and told him what I had done. After presenting all the arguments at his command to dissuade me from going into the army, and finding me still resolute, he said, "Go and do your duty, and if I was as young and strong as you, I would go, too." When my father had gone home, I returned to the recruiting officer and signified my readiness to begin the life of a soldier at once.

Being required to write my age upon the enlistment paper, I wrote "nineteen years." The recruiting officer sarcastically remarked, "Yes, three years ago." And when I assured him of the truthfulness of my statement, he laughed immoderately.

I was then conducted upstairs into the old city hall among those who had been enrolled before me and was introduced as a "new recruit."

The men were all strangers to me, but they gave me three cheers as they also did every new comer. All seemed to realize the fact that we were to pass through our baptism of fire together; and needed this warm assurance of fraternity.

The next five days, we were allowed entire freedom to go and come as we pleased; but none of us got very far from headquarters for fear that something of interest would transpire in our absence. We improved these days in getting as much acquainted as possible with our fellows. The majority were good reliable young men without anything special to bring into notice. And all were full of patriotism and zeal.

One man who gave his name as Wm. H. Osborne, but of whom nothing was known, claimed to have been driven out of New Orleans on account of his Union sentiments. He was of medium size, with red hair and short stiff red mustache and red face. He was very energetic in expressing his determination to fight the rebels as long as he had breath or a drop of blood. A little familiarity with him disclosed the fact that he carried in his boot a bowie knife, long sharp and needle pointed, and apparently was ready to use it whenever himself or friends were insulted. Of course, such a man was very much noticed, and we expected great things of him later.

Another man calling himself James Westbrook was a member of the squad. He was of sandy complexion with thin straggling hair and whiskers. He was very uncommunicative concerning his past; but it was learned that he had been a gambler on the Mississippi. He was first noticed on the streets of Adrian as a well-dressed stranger. He put up at the Cross House that stood where the County Jail now stands. After getting his supper, lodging, and breakfast, he passed out the rear exit through the stable onto the street, leaving his bill unsettled. Next, he was seen dressed as a woodsawer with bucksaw and sawbuck on his shoulder, having pocketed a trifle in money for the exchange of clothing. He soon sold the saw and buck for fifty cents and presented himself for enlistment.

There was one other young man, also of the redheaded persuasion, who attracted some attention at the time. His name was Love Doty, a resident of Adrian and a baker by trade. His common conversation was of killing five rebels before breakfast and of lunching off their bodies while he washed down his repast with their chivalrous blood.

There were some others that made great promises; but the remainder of us who came from the farm and the workshop and knew nothing of war, and little of the ways of the world, naturally expected that such as these would carry off the palm of glory. That they would be found still fighting when we had become worn out and discouraged. We learned long after, that the dog that barks the loudest is not the surest to bite.

While our company was being formed, we were fed at a restaurant and were divided out among the hotels for lodging.

The number now being sufficient to warrant it, we balloted for and elected officers.

April 25 – The officers assumed command, and we were introduced to the foundation of a soldier's life, and the position of a soldier without arms, the facings, and the military salute. There was considerable difficulty experienced in learning us to instantly recognize our right hand from our left; and in learning us to turn on our heels instead of our toes.

April 26 – We were entrusted with guns; and the school of the solider continued. The many spectators always present, though deeply anxious to have us gain proficiency as soon as possible, were afforded more entertainment than they would have been by a circus, in watching our awkward motions. As most of us were more familiar with the spade, the pitchfork, or the axe than we were with the musket, we were as liable to hold our guns in our left hand as in our right; and our feet were always in our way when we attempted to "Order Arms."

These simple drills were to us of profound moment, and we gave them our whole attention.

About this time, to our disappointment, we were informed that the Cadets had been accepted in the 1st Regiment and that we were assigned to the 2nd Regiment. We were very fearful that we would never be sent to war, but our drills were continued with great zeal.

The Cadets, having been ordered to rendezvous at Detroit, Plymouth Church invited both the Cadets and the Guards to attend church in a body Sunday evening, April 28, which both companies did. The sermon was not as interesting to us as it would have been if we had been going with the other company.

After church, a squad of us were taken to the Bracket House and conducted to bed by D. A. Woodbury, afterwards the gallant colonel of the 4th Michigan Infantry.

April 29 – We escorted the Cadets to the depot on their way to Detroit. We returned to our quarters deeply envying them as having been favored in being so soon started for their future field of operation. Our impatience was soon gratified for we soon learned that we would start for the same rendezvous the next day. Many friends came to bid the boys goodbye; and among others my father, sister, Laura S. Haviland, and others came to say farewell to me.

[I always heard that we were related to Laura Haviland (1808-1898). Within the family tree, Harmon's siblings—Almon, Levi, and Mary Jane, all marry Havilands. However, because of Laura Haviland's dedication to humanitarian causes throughout her life, the residents of Adrian, and beyond, always referred to her as "Aunt Laura." This is an excerpt from her obituary:

"Born in Kitley, Leeds county, Ontario, Can., Dec. 5, 1808, she removed with her parents to New York state and in 1826 was married to Charles Haviland. Soon after marriage, they came to Michigan and settled on a farm in the township of Raisin, a few miles northeast of Adrian. No sooner had the Havilands taken up their Michigan residence than they turned their attention to the abolition cause. Being true blue abolitionists they were not slow to show the depth of their feeling against slavery. They became identified with the great body of abolitionists in southern Michigan who assisted the slaves on their way to Canada and freedom. So well known became Mrs. Haviland's sympathies with the Negro and his cause that she was known far and wide as the president of the Underground Railroad."

The obituary refers to her daughter, Mrs. A. Camburn, perhaps Almon Camburn, and her remains were on view at the home of Harmon Camburn, Adrian. Today, a statute of her stands in the center of Adrian, with the inscription, *"A Tribute to a Life Consecrated to the Betterment of Humanity."*]

April 30 — Escorted by the fire department and with firing of cannon, and much shouting by the excited people, we marched to the depot to go by rail to Detroit. Loaded down with many cautions and some good advice, we left the station amid the whistling of steam engines and the ringing of bells, arriving in Detroit safely in the afternoon.

Before leaving Adrian, the old members of the Guard, thinking to make us look something like soldiers, had bestowed upon us their old fatigue uniforms, and we, expecting the State would give us uniforms as soon as we reached Detroit, sent home all our surplus clothing. Our caps were gray, waistcoats light blue and the pants had been originally red, but the material being cheap had faded to a dirty brown color. These being all old and much worn were not very picturesque. At home, where we were well known, this faded uniform was considered a mark of honor, and we boys fresh from school, were naturally pleased with the novelty of our dress.

On our arrival in Detroit, we were marched through the City to the old state fair ground, which was called Cantonment Blair in honor of the governor of the State.

[Governor Austin Blair (1818-1894), served as governor of Michigan from 1861-1864. He's often referred to as the Civil War Governor for his outspoken anti-slavery sentiments.]

Here we were quartered in the sheds left by the agricultural society, which were not ready for us; and the sides were open, and no floors were in some of them. One blanket that was too short at both ends for the smallest man in the company was issued for every two men. The weather that had been growing sour all day, culminated during the night in a cold drizzling rain, which continued the next day. Our thin clothing and little blankets did not protect us from the cold or the rain; and in consequence, we had not much sleep that first night in camp. The rain was a Godsend in one way, if not in another; for from it we were able to perform our ablutions in the ditch beside the racetrack. We were then and there reminded that the State did not furnish towels and a thin pocket-handkerchief was all that most of us had for wipes. Those of us who parted our hair in the middle missed our mirrors and not one in five had a comb.

Workmen were set to work to put our barracks in order, and we, of necessity, had to stay out in the rain. Kindling fires of the best material at hand, we folded our blankets about us and hovered near until smoke, ashes, and dirt had given us and our clothes a coating that completely disguised us against the recognition of friends. Notwithstanding our nondescript dress and vagabondish appearance, our curiosity could not be curbed. We had to take in the town. Soon all parts of the city were infested with these half-civilized-appearing rag muffins, much to the disgust of the citizens, whose upturned noses were plainly to be seen; and they were not at all careful lest we should hear their derisive remarks about these future defenders of the country.

Some kind-hearted and patriotic ladies, hearing that our barracks were not ready and that we were not adequately protected, sent invitations to officers to send squads of men to sleep on their parlor floors. In some cases, these were accepted, but once was enough, and the invitations were not repeated.

Coming as we did from good homes where we had good beds and well-filled tables, there was one other striking contrast for us. The contract to feed the regiment was let to the highest bidder. The contractor of course was patriotic; but he desired to make money. The bakeries and restaurants of the city found ready market for any stale bread they might have on hand. The butter outranked any officer in the state. The finding of a dishcloth in one can of soup and a bar of soap in another was considered as a trifling inadvertence. Our reverence for old age was put to a severe test when we came to eat some of the meat; and the milk had evidently crossed a stream before it reached us. Proper care had not been taken to dissolve the chalk for we certainly found lumps as large as walnuts in the milk. We bore all these trifling annoyances with much less complaint than might be expected under the circumstances.

The companies all having arrived, the regiment was formed and guards posted around the cantonment.

The first morning after the guards were posted, the sentinel at the main gate was approached by a stocky-built dark-browed man, with a stiff black mustache and a strong voice, having a decided nasal twang, who wore a plug hat that had evidently been much used as a chair cushion, and a Prince Albert coat that showed the canvas stiffening through a rent six-inches long in the breast. The guard, seeing his intention of passing insides, said to him, "Halt, you old duffer! You can't go in there."

The dark man replied, "I guess I can go in; I am colonel of the regiment."

The sentinel said to him, "Go way from here. No such old vagrant as you can play that on me."

The dark man replied with some force: "I am J. B. Richardson, Colonel of the 2nd Regiment. **[Most likely this was transcribed incorrectly from the original memoir. Israel B. Richardson was appointed to the 2nd Michigan around this date, so it is most likely the "J" was an "I" in the original.]** Will you call the officer of the guard?" The officer of the guard appeared and admitted the colonel to the camp of his own regiment. But the sentinel that was on duty at the time never liked to hear this story.

May 2 – The colonel called out the regiment and had each company take its place in the line according to designation. In the formation, the Adrian Guards lost their identity as such and became Company D for all future time.

On the next day, our blankets were taken from us and returned with a lining of cotton cloth; and each man had one to himself. Also, we were furnished with broad-soled brogan shoes, preparatory to regular daily drills.

May 4 – The Colonel gave us our first instructions in battalion drill and dress parade. He patiently explained to the officers in the presence of the men what move he desired to make and how to make it. Then came the tug of war for the poor privates. Not being accustomed to the light touch of elbows required of well-drilled soldiers, nor to keep step when we were in line of battle, the long row of men surged heavily toward the center of the line until the men were nearly cursed, and the line was doubled up and broken. Then there was such pushing and pulling, crowding, and squeezing, and treading on heels to regain lost positions that the regiment had to be reformed to straighten out the confusion. This performance was repeated again and again until night released us. There was no trouble to sleep after such a day. Improvement was made from day to day, and we were encouraged.

About this time, orders came from Washington that no more three months' troops would be accepted. The regiment was reorganized for three years by allowing those who were unwilling to enlist for that length of time to withdraw; and those that remained were mustered into the State service for three years to date from the 27 of April 1861. Recruits to fill the vacant ranks poured in, and drilling was vigorously kept up.

[On May 3, 1861, President Lincoln called for an additional 80,000 men. Most likely with this order, came the new requirements for length of commitment.]

May 15 – With something like envy in our breasts, we saw the 1st Regiment fully uniformed and equipped, receive a beautiful flag from the Ladies of Detroit, and start on their way to Washington, while we were yet wearing our blankets in place of overcoats, and perhaps would never leave the state.

As soon as the 1st Regiment was gone, we left Cantonment Blair and took up our quarters in Fort Wayne. Here the drill and discipline were continued; and further advances were made in soldierly qualifications. Also, we were supplied with bright new uniforms, and with them, we felt more like soldiers.

May 25 – The 2nd Regiment was mustered into the United States service for three years or during the war.

Hints now began to be thrown out that we would be soon called to the field. Those who desired to visit home before we moved were permitted to do so.

The sight of a soldier had never been a common one in the rural districts of our state; and the advent of one in full uniform was an event that called together the neighbors. Scores of questions were asked about this one month in camp, and much advice for the future was given. Much of this kindly intended advice was impractical, and I fear most of the remainder was soon forgotten. At least there was not much evidence of its being faithfully followed.

June 6, 1861 – Fort Wayne presented a scene of unusual bustle and excitement. The 2nd Regiment had been ordered to Washington, and preparations were being made to move at once. Officers resplendent in their new uniforms, bedecked with gold braid and accoutered in untarnished equipments, were rushing about looking after the final arrangements. The soldiers, in bright new clothes with shoes polished and brasses shining, were actively engaged in getting everything in readiness for departure.

The last of the preliminary exercises was a march through the principal streets of Detroit and a review by the Governor in front of the Exchange

Hotel. The citizens who had sneered at the half-clad rowdies a month ago, now enthusiastically cheered these fine appearing promising soldiers who were forth to uphold the honor of the state. During the night, the regiment was conveyed by steamer across Lake Erie to Cleveland, Ohio.

June 7 – Landing at an early hour, coffee, bread, and meat were served to the men while standing in the street.

Many citizens of the place came to see the Michigan boys and give them words of encouragement.

A company of juveniles, fully armed and equipped, paid the regiment a visit. The little fellows conducted themselves in true military style and gave the 2nd Michigan three rousing cheers as a send off.

Leaving Cleveland by the Pittsburgh Road, we passed through Hudson, Ravenna, Wellsville, and other places where large crowds of people had congregated to meet and cheer us on our way. At all stopping places, the patriotic Ohio ladies were present with coffee, pies, cakes, sandwiches, lemonade, fruits, bouquets and whatever their loyal hearts suggested would be encouraging to those they regarded as their defenders. A day of excitement among these Ohio towns, relieved by long rides past oil derricks and iron foundries, terminated in the evening at Pittsburgh, Pennsylvania. Here great crowds of people were ready to do anything that seemed necessary, and many things that were unnecessary. Pocket books were open to buy whiskey to fill canteens or bread to fill haversacks. Open generosity was the rule. Real wants were not calculated. To express a wish was to have it gratified on the spot. The need of caution against the universal desire to do something for the boys was very apparent, and the road to Harrisburg was taken with very little delay.

June 8 – This bright and beautiful morning dawned upon us amid the Allegheny Mountains. To boys who had been reared in the comparatively level state of Michigan, the wild and rugged scenery, and the towering hills of Pennsylvania were something to gaze at with awe and wonder. In the presence of these majestic piles of earth and rocks, the individual man shrinks into nothingness, and the immensity of the universe is increased in our estimate.

A little west of Altoona, the railroad runs on the mountainside in the shape of a horseshoe. On the inside of the curve, there is a sheer descent of three hundred feet, while on the outside the rocks rise perpendicular to a great height. Coming upon this spot without warning, with the train rushing along at full speed, one instinctively clutches the hair on his head as if to keep it from flying away.

The ever-varying landscape, as the train sped past lofty mountains, through green valleys and over flashing streams, beguiled the time till Harrisburg was reached in the afternoon.

The Pennsylvania Buck Tail Regiment had a camp here called Camp

Curtain after the governor of the state.

Each member of this regiment wore in his hat the tip of a deer's tail, and was supposed to have killed the deer himself.

At Camp Curtain, our tents were pitched for the first time. Here we spent our first night under canvass. The regiment had brought guns from Michigan, but no cartridge boxes or munitions. Both of these were issued to us here, completing our equipment. Camp duties being all attended to everybody went for a cool bath in the Schuylkill River. Thus refreshed after our long ride in the cars, we slept soundly regardless of our, to us, novel positions.

June 9 – Again taking the rail, Harrisburg was soon left far behind, and the green hills of Maryland appeared shimmering in the sunlight.

The beauty of the scenery along the way banished all thoughts of war until squads of soldiers guarding the bridges recalled the occasion of this excursion.

Baltimore was reached at sunset. A large concourse of people were gathered at the depot, and the fate of the troops that passed through this city a few days before caused a quickening of the pulse and a shade of anxiety, although the crowd appeared quiet and peaceable. Each company left the cars and quietly took its place, and the regiment was dressed in line.

The colonel in his powerful nasal voice gave the command: "Load at will." The assembled throng at once became hushed. They did understand what was to come next. There was nothing meant, only to show that this troop was fully armed, and that no trifling would be tolerated. The line of march was immediately taken up directly through the city. The streets were filled with a dense mass of surging, shouting humanity of both sexes and all ages.

This solid wall of flesh and blood only gave way to the marching troop, step by step, as the advance guard pressed upon them at charge bayonet, and then it seemed to roll up on either side to close sharply in upon the rear of the column as it moved forward. So hemmed in that it would be impossible to do any effective execution in case of attack, with the late occurrences in these same streets fresh in mind and little knowing how soon bricks and stones would begin to descend from the high walls on each side of the street, the mob continuously tongue-lashed with derisive and threatening epithets: "Hurrah for Jeff Davis," "Hang Abe Lincoln," "Northern Mudsills," "Abolition Scum," "Yankee off-scouring," "Nigger Thieves," "Spawn of Hell," and other greetings less refined, the great city of Baltimore gave us that lovely Sabbath evening. To show that there was a little good left in the city, just enough to save it, there was occasionally a faint cheer for Abe Lincoln and the Union. For even with no experience in war, the transition was very rapid from the holiday excursion of the two preceding days to being surrounded by this angry mob who had the disposition, but not the courage, to annihilate these "Yankee hirelings." The regiment moved steadily forward in silence, every man attending strictly to his own business.

The war was very young at this time; and it may be pardoned if the soldiers experienced a feeling of extreme relief when the city was passed in safety, and they were again on the cars for Washington. As the train pulled slowly out of the depot, a rebel rowdy, who could restrain his feelings no longer, threw a large stone through the car window, but hit no one.

Sergeant Kellogg of Company E, who was on the platform, saw the act and shot the man dead on the spot.

Without further incident, Washington was reached at three o'clock in the morning. The regiment was taken into the inauguration ballroom that stood in the courthouse yard, to wait for the dawn of day.

CHAPTER 2 – JUNE 10, 1861

The bright glare of the morning sun aroused us from our hard couch upon the floor before the reveille sounded, in the city of Washington, the capital of the Union, which had so often been described to us as country schoolboys. Few of us ever expected to see the place and fewer still to be numbered with its defenders.

While the officers were engaged during the forenoon with the necessary arrangements for locating a camp for the regiment, the men were at liberty to see the sights.

The public buildings all with marble fronts were the special attraction.

The capitol building was then unfinished. The base of the great dome that now surmounts it was just being laid. The massive stone pillars that now support the long colonnades on the east and west fronts were lying on the streets waiting to be reared to their places. The basement was stored full of barrels of pork, beef, and flour to feed the troops that were expected to arrive from day to day.

Congress was not in session and some thought it a matter of moment to sit in the seat of a senator and write a letter to friends at home.

In the patent office was to be seen models of all the useful inventions and numerous curiosities, but what held the gaze of the young soldiers the longest was a glass case in the center of the great hall containing the military clothes of General Washington, which he had worn, perhaps through more than half the campaign of the Revolution. They are still preserved in almost as perfect a condition as when the illustrious owner hung them for the last time in his wardrobe at Mount Vernon. The coat is of deep blue cloth, faced with buff, with plain gilt buttons. The waistcoat and breeches are of the same kind of buff cloth as the facings of the coat.

In the same case is Washington's dress sword. The handle is iron, colored a pale green, and wound in spiral grooves with small silver wire.

16

The scabbard is of black leather and the belt of whitewash leather, both with silver mountings. Crossing this sword is the staff of Franklin, a long, knotty, crab-tree cane with a golden head, which was bequeathed by Franklin to Washington. This is the "Sword and the Staff" that formed the subject of that beautiful ode by Morris **[George Pope Morris, 1802-1864]** beginning, *"The Sword of the Hero, the Staff of the Sage."* Upon these and other relics of the Father of his Country, the boys gazed with silent awe and admiration.

Visiting the General Post Office, it was a matter of surprise that one could buy a three-cent postage stamp in the imposing and massive structure, as well as at a county post office.

Afternoon, the President desired to see and review this first three years' regiment that had arrived in Washington.

Marching in column by platoon up the circular gravel carriage drive in front of the White House, the President, and General Winfield Scott reviewed the marching troop from the portico.

[General Scott (1786-1866) served on active duty as a general longer than any other man in U.S. history and was known as the Grand Old Man of the Army.]

The two great men, towering to the height of six feet three apiece, were in front, backed by various military officers and government officials. The long gaunt form of Lincoln, dressed in plain black, formed a striking contrast to the portly and majestic figure of General Scott in all the glitter of full military dress.

Returning to the courthouse, our tents were pitched along one side of the capacious yard, the other side being in use as a riding school for cavalry. The instructor of this school was an officer of regular army, and was not at all backward about thumping the knees of the pupils with a heavy stick he carried, if they did not keep their legs in position as he directed. One man fell from his horse and bruised his face quite badly, but this made no impression on the instructor. The man was instantly remounted and the drill went on as before.

June 11 – The day opened intensely hot. A campground had been selected and the first business of the day was to get to it.

Passing out of the city and through the little village of Georgetown on the aqueduct road, a spot was reached about two and a half miles from Georgetown where a clear rocky stream tumbled down to the Potomac. Beside this creek, in an open space on the southern slope of a Maryland hill, camp was pitched and named Camp Winfield Scott.

Immediately back of camp and stretching to the top of the hill, was a growth of small oak timber called in the south "Oak Barrens." Directly in front was the aqueduct that supplies the City of Washington with water. On top of the aqueduct is a good macadamized road. Perhaps forty rods south of the road is the Ohio and Chesapeake Canal, and a short distance further is the Potomac River, at this point about a quarter of a mile wide and clear as crystal. On the opposite shore, which is in Virginia, the banks are high and wooded. It is a beautiful and healthy site for a camp with plenty of good water, but this first march under the fierce rays of a burning southern sun has left the young and tender soldiers but little energy or disposition to explore the surroundings today.

Here in this camp, the 2nd Michigan is to begin in full the life of the soldier and receive the discipline and go through the process of hardening that fits men for active operations in the field of war.

In front, along the entire length of the camp and parallel with the road, canal, and river, is a broad open space, known as the regimental parade ground.

The tents of the companies are set in rows endwise to this parade ground, leaving a street between each company and the next one to it. These streets are called company parade grounds or company streets. Back of the companies stand the company officers' tents and back of these stand the tents of regimental officers: colonel, lieutenant colonel, and major. This is the form of the camp of a regiment in all cases where the nature of the ground will permit. Guards were posted around the camp to prevent ingress or egress to all who had no pass by day or counter-sign by night. Men were detailed to cook such rations as the quartermaster furnished and dealt out to each his proper portion.

The stated roll calls laid down in the tactics were instituted. These were Reveille at sunrise, Retreat at sundown, and Tattoo at nine o'clock p.m. Taps sounded at 9:30 when every light must be out, and everyone not on duty in bed.

Regular hours were designated for company drill in the forenoon and for battalion drill in the afternoon. Those who were sick repaired to the surgeons' tent when the bugle sounded at eight in the morning, and those who were for camp guard went through the formal guard mounting at nine. This was to be the regular routine of our lives during our term of service, unless on the march or in a fight.

To boys reared amid comfort and plenty, with many luxuries thrown in, the following month was full of hardships. Soft still-fed fat pork boiled with a small quantity of rice or beans, coarse baker's bread and coffee, boiled in the same greasy iron kettles after the pork, did not relish after a restless night upon the hard bare ground. Some could not eat the food or drink the coffee, and there was no chance to buy anything else, and very few had money, if there had been a chance. Numbers of the poor boys were from home for the first time and were homesick. Many secret tears were shed in silence by the lads who longed once more to lay their weary heads on downy pillows at home and taste again of Mother's snowy bread spread with delicious, golden butter. How they missed Mother's pies and cream cookies, none can tell so well as those who have suddenly been deprived of them. How intensely home and loving words of home friends were desired none know so well as those who are far from all things like home for the first time. Even written words of cheer failed to reach us. Either those we left behind were neglectful or the mails were tardy in reaching us, but the elastic spirit of youth could not be crushed. These young men had started out to fulfill a purpose and trifles could not turn them aside. Each smothered his own unhappiness as much as possible and strove to encourage others to a cheerful view of the surroundings. Incidents occurred from time to time to rouse the spirits of the men and make the time seem shorter.

The first notable event was the arrest of John Frizelle, or Bull Frizelle as his neighbors called him, a rabid secessionist who resided on the banks of the canal in front of our camp. He was a very large, powerful man and report said had lifted 1,800 pounds. Wonderful stories were told of the fights he had been engaged in and the scars on his face and body seemed to confirm the reports. He was the terror of the country and made dire threats of what he would do to us when he gained his liberty. One day I was detailed with a guard to take this man to his house to get clean underclothes. On arriving at the house, I left the guard outside and entered alone with Frizelle. I was in such fear of the man that I compelled him to change his linen with my bayonet pointed at his heart, although he continually roared at my Yankee impudence. I have often since laughed at my fears. This notorious blowhard was released when we broke camp to join the army for the first battle of Bull Run.

We were fully alive to the fact that we were in an enemy's country and were liable to attack at any time. Exaggerated reports of the nearness and strength of the enemy circulated freely. Sentinels on duty by night vigilant and reported all unusual occurrences. Signal lights were often reported but on investigation proved to be lightening bugs, falling meteors, or other innocent things.

We assisted in building a redoubt at the end of the chain bridge across the Potomac, which was two miles above our camp. This was the first earthwork thrown up in defense of Washington.

Our state sent us each a pair of linen pants and a straw hat. It would make an old soldier laugh to see a regiment in such an outfit. The first rain shrunk the pants to our knees and sent the hats up to a peak. That was the end of this suit.

The army diet was so disagreeable that some risk would be run to get a square homemade meal. With this object in view, a party of four of us surreptitiously obtained the countersign and left the camp late in the evening. Having gone near four miles into the country, we aroused the inmates of a farmhouse at midnight and made some inquiries about the distance to Georgetown and about deserters. Apparently having gained the information desired, we made a request for supper, which the frightened household complied with speedily. A platter of cold boiled beef and cabbage with bread and butter vanished like the dew before the sun. When we were satisfied, we offered to pay but, of course, our host declined, and looked happy that we were about to depart. It was lucky that they would take no money for there was not a cent in the party.

About this time, almost every man drew from the Government a new pair of shoes. The old shoes were pitched into the streets. From throwing shoes at one another in sport, there arose a novel battle between the right wing of the battalion and the left wing, in which the missiles used were shoes. The fight raged for some minutes, and the air was black with shoes most of the time. Finally, the right wing held their fire until all the shoes had been thrown to their side. Then piling the shoes on their left arms, they charged the left, pelting them as they fled. The guards were unable to stay the tide and the pursued rushed out of camp with the pursuers after them. The company officers invoked the colonel's aid to restore order, but he only laughed and said, "Let the boys have their fun." The fight was over, and the shoes out of camp, and order restored itself.

There were other troops in our vicinity. A company of District of Columbia Zouaves guarded the chain bridge. The 1st Massachusetts camped near us soon after our arrival, and about the middle of June, the 3rd Michigan and 12th New York put in an appearance. Of these four regiments, a brigade was formed and the command was given to Col. J. B. Richardson.

June wore away with the same unvarying round of regular military duties, except when the monotony was broken by an accident or incident.

July 4 – The sergeants, by permission of the colonel, conducted the dress parade, and the officers carried muskets in the ranks as private soldiers. Sergeant Wm. McCreery acted as colonel, and ordered the following order read to the regiment, which caused considerable merriment.

Battalion Order No. 1257
Headquarters 2nd Michigan Infantry
Army of the Potomac
July 1861

    1. The orderly sergeants are hereby ordered to the Quartermasters to receive their respective rations of pie, preserves, and cakes, as they will be issued at precisely 5 o'clock every p.m.
    2. Each company will be entitled to three casks of lager beer and fifty bologna sausages every Sunday morning.

Signed by
Bill McCreery, Colonel Commanding Regt.
Augustus Goebel, Adjt.

McCreery afterward became colonel of another regiment, and at the close of the war, was elected to a state office.

July 6 – The 4th Regiment of Michigan Infantry that was rendezvoused by my own home arrived in Washington and camped on Meridian Hill. I hastened to visit them and from them heard the first news from friends at home since I left the state of Michigan.

July 7 – I received the first letters from home friends. This being my first absence of any length from home, the daily recurring changes and the radically different manner of life made the time seem like a year. These anxiously longed for letters made a pleasant break in the tedious monotony of our round of duties. They were read and reread until the contents were memorized.

I shall always remember the terrible accident at this camp. I had not learned the names of the thousand men in our regiment so I cannot give the names, but the circumstances will never be forgotten. The battalion had returned from drill and was dismissed. Two friends stopped to have a little exercise in the manual of arms by themselves. One gave the commands and the other executed. The manual was gone through with the commands, "Ready, Aim, Fire." Suiting the action to the command, the man with the musket, supposing it to be empty, aimed directly at this friend's heart, and pulled the trigger. To his horror and the general consternation of the whole regiment, the gun was discharged, shooting his friend through the heart, instantly killing him. This unfortunate accident cast a gloom over the camp. Here was the first one of our boys killed, and his was the first blood we had seen flow.

Time dragged heavily along, and we were eagerly looking forward to the advent of the Paymaster, while we felt that suitable proficiency in discipline and drill had been made to warrant a movement on the part of the enemy. There was not a thought that when our chance came, anything but victory would be the result.

July 15 – An order was promulgated that all baggage must be securely packed, ready to be deposited with the Quartermaster. Troops to be ready to march at a moment's notice, in light marching order, with three days cooked rations. In "light marching order," no personal baggage is carried except a blanket and rations. Camp rumors were numerous but very unreliable.

All were elated with the hope that we were to strike a decisive blow at the rebellion. In our youthful inexperience, we felt sure of destroying the army that lay between us and Richmond.

July 16 – the Paymaster made his appearance for the first time. It had been most three months since most of us had seen the color of money, and the rejoicing was appropriate to the occasion. We felt that with a little money we could indulge in a little better food on our proposed march through the country. Before three companies had been paid, the order came to move at once.

During a temporary delay, our loaded muskets were stacked. While I was standing in rear of the guns beside Orderly Sergeant Jos. Warrier, a team of mules attached to a heavy wagon ran away. In passing us, the hub of the wagon struck a stack of arms and knocked it down. The sergeant threw out his right hand to catch the guns when one of them went off and shot away three fingers of his hand. This was the first man of our company hurt.

Almost at the instant of this accident, the order to take arms was given and the regiment was quickly on the march. It was three o'clock p.m. as the column moved briskly out of camp and filed up the river. Crossing the chain bridge that spans the Potomac two and a half miles above, we set foot upon Virginia soil for the first time.

Following the main road until nine o'clock at night, Vienna was reached. This seemed to be a point of concentration for here were found troops who had just arrived from different directions. Here the order was issued to bivouac for the night. In the bivouac, the soldier wraps himself in his blanket and lays down with only the vault of heaven for shelter, with his arms and accoutrements by his side. Otherwise, the camp and bivouac are alike.

July 17 – At sunrise, the bugle sounded the reveille. It was a great novelty to see an army rise up from the ground and prepare for breakfast. Looking over a large tract of open ground, the whole surface was alive with human beings. Some were kindling little fires to make their coffee, some were folding their blankets, some were carefully collecting their equipments in a snug pile. Others were lying at full length, sitting, or kneeling upon the ground, and others were walking aimlessly about. With the bright rays of the morning sun glancing over the scene, it was animated and picturesque.

At this place some weeks before, a battery, marked by bushes upon the hill west of the town, opened fire on a trainload of Ohio troops (1st Ohio Infantry, June 17, 1861, Colonel McCook, General Schenck commanding) as they came into the station. The wrecked cars still bore witness to the destruction caused by these rebel cannon.

The life of the little village seemed to be all centered at a small corner grocery. Thinking to gather in a little profit as the tide rolled by, the grocery man opened his shop to the soldiers. Some were buying or pretending to buy, while their comrades were stealing many times as much as was being paid for. About eight o'clock, the army was again on the march with skirmishers in front to prevent a surprise by the enemy. Early in the day, Fairfax Court House was in sight, and the army was put in battle array, but scouts soon discovered that the enemy had retreated. Resuming the march, we came later upon a little village of log houses called Germantown. Many of the houses were on fire and many more were fired by lawless soldiers. Vigorous efforts were put forth by the officers to stop this vandalism, but we did not tarry to see the result.

Pressing on after the retreating foe, we heard the sound of the enemy's guns for the first time, as the enemy scouts fired upon stragglers from our army about sunset. Before darkness enveloped the earth, our army was bivouacked for the night with pickets posted and every precaution taken to prevent surprise.

Wearied with the toilsome day's march, the soldiers early sought rest, and before the night was fairly begun, the stillness of solitary nature rested on the reposing troops.

About midnight, the deathlike stillness was rudely broken by the discharge of musketry, the sounding of bugles, the beating of the long roll upon the drums, the clashing of arms mingled with the shouting of soldiers, and the hoarse commands of officers. Bursting upon the deep stillness without warning, this terrific uproar lasted perhaps five minutes, froze with horror the veins of the soldiers, and shook their frames like ague chill that had but a moment before indulged in dreams of security. From this scene of confusion, order soon came, with every man in his place ready to meet whatever might come. A little investigation disclosed the fact that a body of the enemy's scouts had blundered upon our pickets in the darkness, but were very glad of a chance to get away. The remainder of the night was devoted to rest.

July 18 – The day dawned with a glorious sunshine and balmy atmosphere.

This army, commanded by General [Irwin] McDowell, was early on the march. As it advanced through a fine country, the ground was gradually becoming higher until about noon, it reached the heights of Centerville. A little hamlet with stone church and blacksmith shop stands on this high ground where a road crosses the one on which our army was marching.

From this place, a fine view of the surrounding country is obtained with a distinct sight of the Blue Ridge Mountains in the distance.

When we reached Centerville, we found a deserted camp of the rebels, which had evidently been left in great haste. Camp furniture and cooking utensils had been abandoned. Cooked food ready for eating was left in some instances, and in some others, the noonday meal was yet cooking over the fire. We accepted this as evidence that the boasted southern chivalry was afraid to meet the despised "Yankee Hirelings."

The army bivouacked at Centerville, and Richardson's brigade of Tyler's

Division was to make a reconnaissance toward Blackburn's Ford, which is located where the road from Centerville to Manassas crosses Bull Run.

Moving on the road leading south, we soon came to a stretch of timber. With skirmishers well in advance, we entered the timber where we found the road blockaded in several places with fallen trees. These had to be cleared away to allow the artillery to pass. With small delay, the timber was passed, and we came upon an opening that extended nearly to Bull Run. On the farther bank of the Run, wooded hills arose in terraces to a considerable height. On these hills with the Run between them and us, the rebel army was supposed to be, although not a man was in sight. To induce the enemy to open fire and thus reveal their presence and position, Ayer's battery of twenty pounders was posted on the highest ground in the opening and commenced firing at random. This brought no response until two companies of the 2nd Michigan were thrown forward as skirmishers, supported by the rest of the brigade. As the skirmishers entered the wooded bottom near the Run, the enemy opened a rattling fire of musketry from the opposite bank, and their artillery began a spirited reply to our cannon. A brisk firing was kept up for a few minutes, and we heard for the first time the shrieking of shells and the zip and ping of bullets. Some were killed and a few wounded, but the carnage was not great. (Losses: Union – killed 19, wounded 38; Confederate – killed 15, wounded 53.) We had received our baptism of fire. The plain over which we moved was covered with dewberry or creeping blackberry vines, which were loaded with ripe fruit. Our men picked and ate the berries when not otherwise occupied, with a coolness that indicated no fear of danger. Having learned the position of the enemy and their evident intention to stay there, we were ordered to return to the bivouac of the army at Centerville.

The brigade was commended for good conduct while under fire by General **[Daniel]** Tyler. We were very much surprised ourselves to feel no shock to our nervous systems while under fire or on viewing the dead and wounded.

[General Tyler (1799-1882) blamed for a portion of the Union's loss at the First Battle of Bull Run].

One case attracted special attention. A cannon ball had carried away the lower part of a man's face, making a clean cut in circular form from the upper lip to the neck. This and other casualties were looked upon as the chance of war, and we, only three months from homes where our surrounding influences kept our hearts tender with compassion, were already becoming as stoical as the untutored savages. In our lexicon, there was no such word as fail, and we were sure that we should prove that the boasted southern chivalry was no match for the "Yankee Hirelings."

CHAPTER 3 – JULY 19, 1861

The indications clearly foretold a battle soon to be fought and every soldier had a burning desire to take part in it. When it was announced that our company (D) and Company C would remain at Centerville to guard the baggage and supplies, we were very much disappointed. Although we felt this to be a hardship, we had already learned that a soldier's duty required him to obey orders, whether they be to fight or not to fight. In fact, he must go where he is sent, regardless of consequences.

Accepting the situation with the best possible grace, we kept a keen eye on everything that came within our range of vision. The troops that we had been with, including the remaining companies of our regiment, moved to the front, followed by others that came up from the rear.

All day, we could see more troops arriving who bivouacked in sight of us to the rear and left of town. Mounted officers resplendent in gold braid and shining buttons were riding from command to command in every direction carrying the orders of their superiors.

Citizens in carriages, on horseback, and on foot mingled with the soldiers and came and went as they pleased.

The civilians all seemed to be well posted as to future movements of our army. In exchange for what we could tell of the skirmish at Blackburn's Ford, they gave us their understanding about the coming battle. In substance, it was about thus: It was thought that our troops would all be up by Saturday night, ready for an attack early Sunday morning. General **[Robert]** Patterson, in command of a division of our army at Winchester, was watching the forces of the rebel General **[Joseph E.]** Johnstone. If Johnstone attempted to come to the aid of **[General P.G.T.]** Beauregard at Bull Run, Patterson was to attack him or come down on the right flank of the rebels in our front and help to destroy it.

[General Beauregard (1818-1893) Confederate soldiers win the

First Battle of Bull Run, which is also called the First Manassas.]

In our ignorance, it seemed all right then that these things were being discussed, but afterwards we learned that it was all wrong.

July 20 – I was posted with a detail of four men on a high point commanding a view to the southeast. We could see mules over and beyond the troops in that direction. My instructions were to report to my commanding officer any suspicious movements that I observed in that direction. As three days in the field had not learned me very much of warfare, I was not very clear in my own mind what constituted a "suspicious movement," and the officer that gave me my instructions could not enlighten me much.

But with five pairs of bright eyes in as many wide-awake youthful heads, I felt that we were competent to detect anything suspicious. With one to keep close watch of the country to the rear of our army, the others were at liberty to gaze on the lovely landscape, made more beautiful in the foreground by the many colored uniforms of the army and the glint of the sun upon their burnished arms. The various regiments, each a thousand strong, some with fancy names and fancier uniforms, began to move to their places in the line for tomorrow's battle. As they pass our post of observation, let us take a hasty note of their dress.

As they march gaily forward with bands playing and banners flying, the 1st Michigan appears in dark blue fatigue caps, jackets, and pantaloons, while the 3rd Michigan is dressed throughout in gray.

Ellsworth's Fire Zouaves wear red fatigue caps, braided with gold, Zouave jackets braided with yellow, cut away in front to expose an embroidered dark shirtfront, orange sash, red pants, and buff leggings. Billy Wilson's Zouaves wore upon their head the Turkish fez with tassel and a white turban, Zouave jacket and shirt, scarlet sash around the waist, Turkish pants of red, bagging at the knees, and white leggings. Hawkins Zouaves wore felt hats with yellow cord, jackets dark brown corded with yellow and red pants. Garibaldi Guards wore Garibaldi hats and jackets with dark blue pants. The Mozart Regiment wore felt hats and dark blue clothes trimmed with red cord. The 69th Highlanders wore Scotch caps and Scotch plaid with kilts. The Hussars were dressed in light blue, the caps and clothing trimmed with white braid.

There was also a regiment of riflemen dressed in green with frock coats.

These brilliant variations mixed in with troops that wore the regulation uniforms made a splendid pageant. They were moving up to join battle with the Louisiana Tigers, Texas Rangers, Black Horse Cavalry, and other fancy rebel regiments.

After tomorrow's encounter, both Federal and Confederate will drop their fancy titles and assume their plain state number as their proper designation.

July 21 – Early in the morning, the booming of cannon and the rattle of musketry was heard along Bull Run. A little later the cannonading became terrific, and the discharge of musketry was one continuous roar.

Just at Centerville, rises one of those abrupt elevations called a knob. From the top of this, we could see the smoke of the guns, and away in the distance, could be seen with the naked eye, signs of animated life in one of the gaps of the Blue Ridge Mountains. The glass revealed this to be a large body of men defiling down the mountain.

In confirmation of the reports of yesterday, we believed these to be Patterson's men coming to take the rebels on the flank. While the contest was raging hotly in front, a battery of three-months' men from New York, whose time had expired, came back from the front, deliberately unhitched their horses from their guns until the day was decided. They marched away for home, leaving others to fight the battle, and the sound of strife behind them seemed to quicken their footsteps.

At ten o'clock, we were ordered to join our regiment. We found them on the Blackburn Ford Road, near the same ground we had occupied on the July 18, our brigade being the extreme left of our army. While we could hear the thunder of artillery, the roll of musketry, and the cheers from the desperate struggle going on to the right of us, there was but little of the action in our sight. But we watched that little action with great interest. On a knoll in the open field, one of our batteries was shelling the high ground across the Run occupied by the rebels. Once in a while, there would be a little commotion, showing that some execution had been done. They replied but seldom, and there the damage was chiefly to the treetops.

In the afternoon, there was indication that the forces in our front were being withdrawn. Then there was a lull in our part of the field, but the main battle was going on vigorously. About five o'clock, a heavy body of troops was seen opposite us across the Run. From their movements, it was thought they would try to turn our left. At once, active preparations were made to move with speed when the time should come.

Richardson, being in command of the brigade, our regiment was in charge of Lieutenant Colonel Williams. He ordered us to lay off our blankets and jackets. It being hot weather none of us wore vests, and this left us with only shirt and pants. Soon the order came to move to the left on the old Union Road at double quick. The movement continued for a long distance until we came into an open field just as the rebels emerged from the woods opposite. Closing up our ranks, we at once advanced with a tremendous yell. The rebels, thinking they had encountered a large force, retired before we had a chance to fire a shot. A young aide-de-camp galloped along our line and excitedly told us we had saved the army. "Michigan would be proud of us," and "Everybody would love us." As the sounds of strife could still be heard, we knew that the rebels were not all whipped. If another yell would decide the day in our favor, we were ready to give it.

Night was rapidly closing in and the firing had almost ceased when we were ordered to a cornfield to our left and ordered to sleep on our arms. We did not know that the groups we had seen coming down the mountain in the morning was Kirby Smith's division of the rebel General Johnstone's army instead of our own men under Patterson. They had left a small force to beguile Patterson, while the others stole away to assist Beauregard at Bull Run. That was the cause of the rebel advance after a seeming retreat.

We knew nothing of the situation but expected to renew the fight at daybreak in the morning. The blankets and jackets we laid off earlier in the evening were now sadly missed as the cold dews of night fell upon us. We also began to suspect that we had donated them to our enemies. Gathering a small handful each of the green cornstalks, we stretched our bodies upon them to shiver the cheerless night away in vain attempt to get a little sleep and refreshing rest to brace us up for the arduous duties of the morrow.

July 22 – At two o'clock in the morning, we were quietly aroused from our comfortless beds and bidden in low tones to get our place in ranks with as little noise as possible. Supposing we were going to take our place in line to renew the battle of yesterday, there was no delay, and soon we were moving briskly toward Centerville. On reaching that place, we were surprised to see no sign of troops, and when the head of our column was directed to the rear, we were greatly astonished.

Then for the first time, the idea flashed upon us that the battle was lost to us and that we were the rear guard of our army.

[The First Battle of Bull Run on July 21, 1861, resulted in a Confederate victory. General Robert Patterson's inability to hold Johnstone's small group of Confederate soldiers led to the Union's defeat and Patterson's end to military life. Union troops retreated back to Washington, and Lincoln realized the war would be a long one. The battle resulted in 4,878 casualties out of 60,680 soldiers. Union: killed 460; wounded 1,124; 1,312 missing and captured. Confederate: killed 387; 1,582 wounded; 13 missing and captured.]

As daylight dawned, the disaster to our arms became more apparent in the waste and destruction that impeded our march. The panic-stricken teamsters in their frenzy to get away from an unseen enemy had furiously driven their teams toward Washington, unheeding and little caring whom they crashed against in their wild flight. For miles, the road was half blocked by broken government wagons, overturned gun carriages, smashed hacks, and demolished buggies. Camp furniture, broken guns, and accoutrements strewed the ground everywhere. All the paraphernalia of war was scattered in one grand wreck from Centerville to Fairfax. Evidently, in some instances, where wagons became disabled the drivers used their knives to cut the mules loose and continued their flight on their backs.

The whole scene indicated a disastrous route to the proud and confident army that had but recently advanced over this same road.

As we slowly followed our retreating army, rational thoughts began to shape themselves and take lodgment in our brains. We began to realize the foe was our equal. The hope of putting down this rebellion in ninety days seemed doubtful. We began to look forward to a long, hard, and bloody struggle. That our government would continue the struggle and ultimately triumph, we had no doubt.

With deliberation, our ranks in good order, the march was continued to Fairfax where a short halt was made to eat the remnants of our rations. With little delay, the march was resumed, and about noon, the rain began to fall. This involuntary bath was very unwelcome to us of the second regiment in our shirtsleeves with no protection of any kind. The tramp of horses and thousands of men in advance of us soon punched these clay roads of Virginia into a bed of mortar. Wallowing in the slush and mire, disappointed, chilled, and gloomy, we struggled on, most of the men keeping their places in ranks, but many, unable to bear the strain longer, straggled from their places and each on his own hook, pressed for the Potomac River. The sight of these men streaming along without order, through the mud, with the rain falling copiously, each carrying his arms in any position that suited his will, with sullen brow and dejected mien, produced an impression that can never be forgotten, but the recollection produces no pleasure. About four o'clock, wet, tired, hungry, worn out and discouraged, from Arlington Heights, we beheld the Potomac at our feet. No doubt, the sight was welcome, but the soldiers made no expression. Their condition was too desperate, their feelings too deep for outward expression. Those of the 2nd, who had held out to the last, were conducted to a barn on the flat near the river, where we found shelter from the pouring rain.

For myself, I thought not of food or danger. Climbing to the haymow, I rolled up a great flake of hay, snuggled into the hollow and let the hay roll back upon me. Finally, I was roused by somebody walking over me. Crawling out of my nest, I found the bright sun of another day shining high in the heavens, and my clothes were as dry as powder.

The road we had followed from Fairfax led past a dirt fort on Arlington Heights, called Fort Albany, down through the flats and across the long bridge into Washington. Near the base of the Heights, the road from Georgetown to Alexandria passed through the Arlington estate and crossed the road to Fairfax at right angles. Half a mile to the west of this road crossing is Arlington House, the property of Robert E. Lee at the outbreak of the rebellion. Across the river, the City of Washington was spread out like a beautiful panorama.

July 23 – Early the stragglers came in and soon the regiment was complete. Rations were issued, and we moved out into the beautiful meadow below Arlington House to bivouac until our tents and baggage could be brought from Camp Scott. While on the meadow, Senator Chandler of Michigan came to see us and made a speech full of praise of our conduct during the late disaster. Among other things he said, "If the Black Horse Cavalry had charged on you, I know you would have buried them." At the close of his speech, there was an attempt to give him three cheers, but our spirits were too utterly crushed to bring forth a cheer. Not a cheer could be wrung from the depressed hearts of the boys who had followed in the wake of, and covered the retreat of, a demoralized army.

When our tents arrived, we pitched our camp on the Heights, opposite Fort Albany, in a field having an octagon house in one corner. The house was used as regimental headquarters. A house at the cross roads was brigade headquarters. And Arlington House was occupied by General Sanford as Division headquarters, while Army headquarters were in Washington. This camp was to be our home for three months. Our occupation was the drill, the picket, and work on fortifications for the defense of Washington.

July 25 – General George B. McClellan formally took command of the Army of the Potomac.

[After the First Battle of Bull Run, when President Lincoln realized the war would be long and costly, he removed General Irvin McDowell and replaced him with Major General George B. McClellan.]

The demoralized mob that returned from Bull Run was to be thoroughly organized and disciplined. At the same time, a chain of forts was to be built on the Virginia side from the Chain Bridge above Washington, along Arlington Heights to Alexandria, a distance of about twenty-five miles, ending with Fort Lyon overlooking and commanding Alexandria.

A great portion of the country was covered with oak barrens. This timber had to be cleared away to prevent its sheltering an advancing enemy and to give range for the artillery. There was no danger of an idle summer, although a few days for rest and recuperation of spirits was allowed us. The spirits of youth are ever buoyant and many an incident helped to beguile the intermissions of labor.

July 26 – Procuring a pass to go to Washington, I visited the Smithsonian Institute. A delightful day was passed in looking over Wright's gallery of Indian painting, fossil remains, mineral specimens, preserved bugs, fish and snakes, quaint arms, and implements of war presented to our government by savage nations and many other curiosities. But that which caused me to linger longest was the great-corrugated lens in the lecture room. Looked at through this lens, the eye appears larger than the human head.

July 28 – Sunday. The regulations require each commander of a regiment to inspect his command every Sunday morning. Every man is expected to be fully uniformed, armed, and equipped. Every strap, brass, and button must be in its place. Forty rounds of ammunition in the cartridge box and the knapsack with blanket rolled and strapped to the top must contain a change of shirts, socks, and underclothing. The movements for inspection are very formal, and when the proper position is reached, the soldier unslings his knapsack, opens it, and lays it on the ground before him. Then he stands perfectly motionless, looking square to the front until the inspection is over. The officer passes slowly along the front of each rank, then in rear, finally taking a peep into each knapsack, carefully noting if everything is as it should be, until all have been inspected.

Sylvester Larned, a man of much ability as a lawyer, but with absolutely no military experience, had just been commissioned lieutenant colonel of our regiment. Richardson being in command of the brigade, it fell to Larned to conduct this first inspection after the **[first]** battle of Bull Run. With the commands written in successive order on a scrap of paper, he proceeded with an air of great authority until he had the regiment in column by company in close order. Then he accidentally skipped the order to open ranks and gave the command to unsling knapsacks. The men stood like blocks of stone, paying no heed to the command. Repeating the order with a great show of authority and louder, he was surprised that the men made no move. Turning to the adjutant, he asked, "Why don't the men obey my commands?" The adjutant informed and reminded him that he had failed to give that order. Correcting his error, he at length got the men in proper position, and began the inspection, somewhat irritated by his own blunder. Being one of the first men inspected, I was severely reprimanded for being in my shirtsleeves and without a blanket. I informed him that I left them at Bull Run by command of Major Williams and had been unable to get new ones from the quartermaster. Having had no pay since I was in the service, I could not buy. I was sharply commanded not to talk back and ordered never to appear again in inspection without a uniform. If he had remained with us until 1863, he would have inspected men with less clothing than I had. Idleness breeds mischief in the army as elsewhere.

August 3 – Sergeant Kellogg, who had some experience in warfare, gained in Florida among the Indians, proposed to the non-commissioned officers that we get up a private expedition to go and hunt up the enemy. Raising a party of twenty with Kellogg as captain, we stole out of camp. Sergeant A. Goebel was chosen lieutenant and myself 2nd lieutenant of the company.

Going directly to the outpost on the Orange and Alexandria railroad, we told Colonel Kerrigan, the commander, truthfully that we were on our own responsibility and desired to go out and stir up the enemy. Promising not to tell how we got out, he gave us the countersign and a guide and passed us outside the lines. Crossing the country to the southwest, we came in sight of the spire at Falls Church. Deploying as skirmishers, we moved cautiously forward until the glisten of bayonets could be seen where the rebels were on picket. Halting at a house that was in the midst of an orchard, Goebel was left in charge while Kellogg and I crept forward along a cornfield fence. Arriving at the opposite side, we peeked through the fence into a pasture lot and disconcerted the rebels advancing on us in good order from the opposite side of the field. We paused just long enough to estimate their number at two hundred, and then hastily rejoined our comrades. Informing them of our discovery, we silently and swiftly transferred ourselves to the next woods beyond the opening, where he held a short council of war at which it was decided that we just escaped being made prisoners, and that we had urgent business in camp.

Taking our way without delay in that direction, we found upon our arrival an order had been issued for the arrest of the whole party. Kellogg went to Richardson and assumed the whole responsibility, and upon assurance of the general that we should only be censured, gave him a list of the party.

August 4 – An order was read at dress parade that was full of praise, all but the last sentence. In that sentence, the general said he felt called upon to censure us for leaving camp without leave, but was sure it would not be repeated.

For the next month, the regular routine of drill, chop trees, and work in the trenches was established as soon as we had our coats and blankets. An occasional incident was all that broke the monotony.

Colonel **[Israel B.]** Richardson, when he took command, was a bachelor. He had taken advantage of the time to go to Michigan and bring back a young wife. One morning, with his wife and a basket that would hold about two quarts, he took a walk to a farmhouse near our camp where there was a peach orchard, loaded with fine ripe fruit. Getting his basket full of peaches, he politely asked how much to pay. The reply being "fifty cents" he paid the money. As he had posted a guard over the orchard to prevent the peaches from being stolen, he now ordered the guard to return to camp. As he returned to camp, he said to every soldier he met, "The guard has been taken off that peach orchard." Before sunset, there was not a peach left.

Milk was a luxury hard to get in the vicinity of a large army. Two of our boys found a man with milk and seeing a chance for profit, they engaged all he had for an indefinite period. Bringing this to camp to sell every morning, they crossed a creek, the water of which had a strong mineral taste. The customers complained bitterly that the milk tasted like creek water. They were assured that it was just as it came from the cow. This went on a few days until a fish three inches long was found in the milk. Then the colonel interfered and prevented them from selling any more milk in camp.

About this time, we began to drill by brigade. This was a novelty at first, but we soon learned that the evolutions required many long marches.

The 12th New York Infantry was in our brigade. By some misunderstanding, they claimed that their term of enlistment had expired and they refused to drill or do any other duty. In fact, they had mutinied. Colonel Richardson gave them a peremptory order to come out for drill, and they refused to obey. Starting out with the 2nd Michigan, ostensibly to drill, when passing the camp of the 12th, he quickly changed the direction, drew up directly facing their parade ground, and commanded them to fall in for drill. Not a man of the 12th made a move to obey. Ordering the 2nd to load at will, he called out in his powerful nasal voice, "I will give you just five minutes to get your places in ranks." As the steel rammers of the 2nd rang in the gun barrels sending home the ball cartridges, there was great haste in the 12th to form the regiment inside the allotted five minutes. They went out and drilled like men and this ended their mutiny.

The last of August, our brigade was formally reviewed by General George B. McClellan, accompanied by President Lincoln. This was an honor not accorded us or many other brigades later on in the war.

The monotonous drill, dig, and chop went on into September with no notable break.

September 6 – Company D was detailed to go out on the Fairfax Road about two miles to Hunter's Chapel and construct a rifle pit. Hunter's Chapel is a small frame church on the Hunter estate. The Hunter mansion being empty, we occupied it while at work there.

The front part of the house is a square structure three stories high with a wide hall through the middle of each story, with two square rooms on each side of the hall, making in all twelve square rooms.

Back is a wing two stories high with kitchen and dining room in the lower story and servants quarters in the upper. Surrounding the house is a broad level plantation with an immense peach orchard. Before the place was deserted, a fine vegetable garden, a large patch of sweet potatoes, and a field of lima beans had been planted. These were just in their prime, and we used them to supplement our government rations during the two weeks of our stay.

The peach trees were cut down and the top end of the limbs sharpened. They were then placed in a row from the chapel across the road to a stone barn eighty rods to the south with the sharpened tops bristling toward the enemy. Then digging a wide ditch the entire length, we threw the dirt upon the stems of the trees, securely holding them in place and at the same time forming a dirt breastwork for infantry. An enemy approaching this work would be exposed to fire a long distance and would be impeded by the sharp limbs when close to it. While at this work, I was called up by the captain at midnight one dark night and was ordered to pick four men to go with me on special service.

The captain took us about a mile in front to the picket line of the London and Alexandria Railroad. Relieving a post of the regular picket, he posted us at a point where a stream of water noisily tumbles through a culvert under the railroad. Our instructions were to capture or kill any living thing that attempted to pass through the lines at that point. We kept a sharp lookout but nothing appeared to us, and we are yet in ignorance as to what we were expected to capture or kill.

September 19 – We had our rifle pit built and the slopes nicely sodded when the 5th Michigan appeared and relieved us, and we had to relinquish our fine house and fat living and rejoin our regiment at Camp Arlington. About this time, **[Israel B.]** Richardson was made Brigadier General and Lieutenant A.M. Poe **[This is most likely Orlando M. Poe, which suggests an error in transcription from original]** of the United States Engineer Corps was appointed our colonel.

The easy-going Richardson had no camp guard except over the quartermaster's stores, and discipline was very loose. Poe at once put ninety men on camp guard and took every means to raise the standard of discipline in the regiment. We soon learned that it was best to obey our young commander and soon he had a regiment that he was proud of, and we had a colonel that we trusted and respected.

September 20 – The regiment went out five miles on the Fairfax Road to Bailey's Crossroads to do picket duty. The Crossroads has a church, blacksmith shop, store, and three or four dwellings. At this place, the Fairfax Road is crossed by another that passes over Munson's Hill about a mile to the north, which was crowned with rebel batteries. Between the buildings and the Hill was the Bailey plantation with its buildings in the middle, half a mile to the northeast. The church was used as our headquarters and from here, pickets were sent out.

The ground between the Crossroads and the base of Munson's Hill was comparatively level, and the road was in plain view all the way to the top of the hill. Our picket line and that of the rebels were about forty rods with the Bailey house between. In our ignorance of legitimate warfare, we supposed we must shoot at a rebel whenever we saw him. This led to many useless casualties on both sides. Our first experience on a picket found us with most of our company in a piece of timber near the Bailey house with one post in the open field in front of the house, protected by a pile of rails. Our officers left us for the night and went back to sleep in the church. Just after daybreak, we saw a force of rebels back of Bailey's house, and we jumped to the conclusion that they were advancing to attack us. We opened fire upon them, and they returned the fire with spirit. Not knowing what else to do, we hastily retreated. The line of retreat of the post behind the rails lay back over the hill in plain sight of the rebels. Doty, the redheaded individual who was going to eat rebels for breakfast when he first enlisted, was one of the men on that post. The rebels gave them a parting shot as they went over the hill. From my security in the woods, I heard an unearthly howl from the opening and quickly looking in that direction, I saw Doty at least four feet from the ground with legs and arms extended, his musket several feet high in the air. The picture, accompanied by the despairing yell, was enough to provoke a smile from a dying man. Upon reaching a place of safety, it was found that the part of Doty's anatomy upon which he would naturally sit had a blister half an inch wide and three inches long as if a small firebrand had drawn across it. Other injuries he had none. These were the last rebels Doty ever saw, unless he saw them in Canada, for he deserted and fled to that country before he had another chance. The officers soon came with more men and restored the line when we found the rebels were only relieving their pickets.

One afternoon while off duty, I was standing with others in front of the church when Colonel Kerrigan of New York rode up in a state of extreme intoxication. At the same time there was a rebel officer riding down Munson's Hill. The distance between the two officers was three quarters of a mile. Kerrigan slid off his horse and demanded a musket. Asking one of the men to hold his horse, he leaned against the horse, took deliberate aim at the rebel officer, and fired. To our surprise, the rebel dropped from his horse and it galloped away rider less. Kerrigan was powerless to remount and after being helped on, he gave the man who held his horse a two-dollar and a half-gold piece, then rode away.

The officers did not stay upon either the Union or Confederate picket lines, and it finally dawned upon the privates it was better to talk with one another than to be shooting all the time. When either party wanted a parley, they hoisted a white handkerchief on a ramrod. When this flag of truce was seen, some of the men came out between the lines from each side and had a chat. Most of the men had been out to talk, and one day when I was on duty near the road, a man came out with a white rag. I thought this was my chance to talk with the rebels. Going one third of the way and the reb coming about the same distance, he told me he dare not come any farther, and he wished to send a letter north to friends. Telling him I would come no more than half way, after a lengthy parley, I turned to go back when his friend opened fire upon me. Being in plain sight with nothing to hide behind, I knew it was useless to run or dodge. Lifting my hat to them, I deliberately walked back to our line, a scared, but a wiser man. No rebel dared to show his head the rest of that day and parleying came to an end.

September 27 – We returned to Camp Arlington.

September 28 – The left wing of our army made an advance, and the rebels evacuated Munson's Hill.

The next two weeks, we took up the old monotony of alternate drill and work.

Officers sent out to fill vacancies often knew little of military. One of these violated several military usages when at midnight, he withdrew the camp guard and posted them as sentinels along the road between Camp Arlington and the outpost, to stand all night. His instructions were to demand the countersign at two hundred paces. Upon his return along the line, we distinctly heard him give the word "Toronto" sixty times.

As the countersign is expected to be given in a low voice over the point of the sentinel's bayonet, this officer had to be regulated in the morning.

CHAPTER 4 – OCTOBER 12, 1861

Troops were continually arriving from the north and the vast army that was concentrating required more room for camp. The older regiments were pushed farther out toward the enemy, and their places filled by the fresh arrivals.

We were gratified by the intelligence that we had marching orders. "Leave, not to return." Taking the road that runs parallel with the Potomac, we followed it down the river to Alexandria, about eight miles from Arlington. From glimpses as we passed through Alexandria, it appeared to be a city of about twelve thousand inhabitants.

There were many good buildings, but wherever these were, there was sure to be old tumbledown shanties surrounding them. The streets were one continuous mud hole and the sidewalks were broken and in a generally wretched condition. Nearly all the business places were closed or used as quarters for the soldiers who guarded the place.

The people on the streets were Negroes and soldiers. Altogether, the place presented a scene of desolation. This was very different from the ideas our school histories gave of the chief shipping port of Virginia.

Passing through Alexandria about two miles, our camp was made on Eagle Hill near Hunting Creek, on part of George Washington's former estate. Our tents were pitched on an elevation that commanded a view of the Potomac and Alexandria, while Washington could be plainly seen in the distance. Here we were to resume our old life with pick, shovel, and axe with the exception of an occasional reconnaissance in which we marched out toward the enemy and then marched back again, and once in two weeks forty-eight hours of picket duty five miles out on some part of the line. About this time, our pickets were pushed out so as to form a continuous line from the upper Potomac near Leesburg to the lower Potomac below Alexandria.

October 20 – The 2nd Michigan, being detailed for picket duty, was posted five miles out on Rose Hill Plantation, owned by Colonel Mason. Dogue Creek traverses this estate and the bottomland, a mile wide, is a beautiful valley enclosed on either side by hills, upon which the oak, chestnut, and birch interspersed with scrub pines form a pleasing and variegated foliage. Rose Hill mansion stands on a hill on the north side of the valley. The face of the hill to the south, artificially terraced so that five or six levels rise one above another, presented a scene to us by the creek that was rarely beautiful.

October 24 – The order was promulgated that our regiments were assigned to the division of General **[Samuel Peter]** Heintzelman, and we reported to him at Fort Lyon the same day. Fort Lyon was an unfinished dirt fort at the extreme end of Arlington Heights overlooking Alexandria. The view from t his point was comprehensive. It took in Washington, the hills on the Maryland side, miles of the Potomac River, a large surface of country to the south and a wide valley to the northwest that gradually grew higher and narrower as you approached the head where stands Fairfax Seminary seven miles from the fort.

Standing one day on the ramparts of the fort, I witnessed the sending of messages by means of signal flags around a twenty-five mile circuit with only the four following stations: Fort Lyon, Fairfax Seminary, Capitol at Washington, Lunatic Asylum on Maryland Hills, and back to place of beginning, and I could see the yard square flags at all of the stations with the naked eye. This side view was ours from Camp Lyon, and we were here to help finish the fort, clear away any timber that might shelter an enemy and to do our portion of duty on the distant picket line.

Sunday, October 27 – A number of boys in Company D were desirous of attending church in Alexandria. Fearing there was a greater desire to get into the city than to worship, and that the men would commit some excess while there, the Captain offered to give me a pass for all the men I would be responsible for, all to have their shoes blacked and wear white gloves and side arms in good order. About twenty accepted, and I conducted them to Christ's Church, or what was known there as Washington's church, from the fact that Washington always attended this church after the revolution and paid the highest price for a pew. Some of the Washingtons had ever since occupied the same pew. The church is of brick with a square hip roof and a belfry at one corner. It was finished in the year 1773. Climbing ivy nearly covers the walls, giving it a very pretty rustic appearance. The inside is furnished with box pews of dark wood, both on the floor and in the gallery. The services were conducted according to the ritual of the Episcopal Church. The congregation was composed of about four hundred soldiers and one hundred civilians. In looking over the audience, I saw many of the hardest characters in the army, yet I did not notice one indecorous or improper move or act. To all appearances, this congregation was as devout as those we see in times of peace.

October 28 – A bright and balmy day like the early autumn in Michigan. Just the day to stroll away from the earthly cares and dream amid the falling leaves. With a comrade, I obtained leave to visit Mt. Vernon, the home of Washington. The distance from our Camp was eight miles. Traversing the meadows and patches of woodland and leaping the little streams in our full tide of youthful vigor, the walk itself was a rare pleasure. As we neared Mt. Vernon, the first object to attract attention was Washington's circular barn, which stands a mile or two north of the mansion. The lower half of the wall was of brick in tolerable good condition, but the upper, wooden part was sadly dilapidated, and the roof and sides were falling to pieces from decay. From the barn, we went on through a piece of timber and across some fields to the mansion. In a lovely spot on the brow of a gentle slope nearly a hundred feet above the water stands the mansion with its accessories. The lawn extends to the abrupt river bank and in front the Potomac sweeps in a magnificent curve, while beyond are the green fields and wooded hills on the Maryland shore. The building is of the most substantial framework, two stories high, ninety-six feet in length, thirty feet in depth, with a piazza fifteen feet wide extending along the entire eastern or riverfront, supported by sixteen columns twenty-five feet in height. Over the piazza is a light balustrade, and in the center of the roof is an observatory with a spire. Along the roof on either side are dormer windows, near the western corners are the kitchen on one side and the laundry on the other. These are connected with the dwelling by open colonnades, each with roof and pavement. Inside, the hall running through from east to west and the rooms, are wainscoted and corniced in such a manner to convey great solidity. In the large drawing room is a beautiful chimney piece of white marble, which was presented to Washington by Mr. Vaughn of England. The mantle is highly ornamented, and the three tablets that form the frieze under the mantle present domestic agricultural scenes exquisitely wrought in high relief. The hearth, also of white marble, is inlaid with ornaments of highly polished light-brown marble. The work was done in Italy.

In some of the rooms were relics of Washington. His packsaddle, leather camp chest, mirrors, vases, pictures, and the bedstead on which he died. Our limit of time did not permit us to go into minute details concerning all there was to see.

In the edge of the grove at the right of the lawn and near the riverbank is the old vault where the remains of Washington were first entombed and remained for thirty years. When he visited American in 1834, Lafayette went down alone into this vault to commune in spirit with his dear old friend.

At the foot of a hill on the edge of a wooded glen, through which a tiny brook gurgles, southwest of the house, a new and better vault has been built according to directions left by Washington. The vault, which is of brick, is entirely overgrown with vines. In front of the vault is a vestibule with an iron picket gate. Within the vestibule stands a marble sarcophagus upon the lid of which is sculptured an American shield over the flag of the Union, surmounted by an eagle. This contains the remains of Washington. By its side is another sarcophagus, perfectly plain, which holds the remains of Mrs. Washington. On the east side of the vault and in front, rest the remains of several of the family under marble monuments.

An atmosphere of peace and rest seems to pervade every part of the premises, and standing here at this great man's tomb, a meek and lowly spirit appears to descend and take entire possession of one's being. Though surrounded by hostile armies, this venerated spot was never desecrated by armed strife or partisan brawls. Here Federal and Confederate, if they meet, are at peace.

As the day drew to a close, we hastily walked through the garden and conservatory and took our way back to camp, feeling that the pleasures of the day had been unusually full.

From a scene emblematic of peace, we turned to all the pomp of war.

October 29 – An extract from Milton aptly describes the morning and the display of troops this day on the occasion of McClellan's grand review at Bailey's Crossroads.

"Now went forth the morn, such as in highest heaven, arrayed in gold, Emphyreal; from before her vanished night, shot through with orient beams; when all the plain covered with thick embattled squadrons, bright chariots and flaming arms, with fiery steeds, reflected blaze on blaze."

With the first light of day, all the camp along the Potomac was astir. The muster place was a little valley whose sides gently slope toward the center, in the center of the Bailey Plantation near Munson's Hill. Early in the morning, eighty thousand men in bright uniforms with shining brasses and burnished weapons were massed compactly in serried ranks on the slopes facing toward the middle of the valley.

On the same field were more than fifteen thousand horses and over two hundred cannon with attendant caissons. This was the greatest number of troops ever assembled on one field at the same time in America, and the nature of the ground was such that all parts of the field could be seen from any other part.

This vast array was arranged with exact precision, its divisions, brigade, and regiments each in its proper place, with the due proportion of artillery and cavalry attached to each division. This nightly host was placed in position with very little bustle or commotion.

All being in readiness, every man became perfectly motionless. Not a move could be seen. Not a sound broke the perfect stillness until General

McClellan and his brilliantly mounted staff rode upon the field accompanied by President Lincoln and his suite. At this first sight of this glittering cavalcade, the cannons boomed a national salute. As soon as the deafening thunder died away, the President and the General, followed by all their train, rode around the valley in front of the line, and viewed the troops. As each successive division was reached, the drums rolled, flags dipped, and the soldiers presented arms. As soon as the circuit of the field had been made, the President and the General took their stand in a convenient position, and the troops passed in review before them. The Infantry moved in "double column closed en masse," that is, one hundred men abreast with each succeeding rank as close to the preceding one as men can walk with freedom.

The Cavalry were fifty men abreast and the Artillery six guns abreast. The time required for this dense column to pass the reviewers stand was three hours and a half. Each regiment passed off the field directly to its own camp, which was from one to twelve miles distant.

To have witnessed this magnificent pageant and to have been one of its units was a great event in my young life.

In returning to camp from the review, our route led us by Fairfax Seminary. The building is an imposing one with a very tall spire. It stands in a commanding position on high ground at the head of an alley that leads down to the southeast. At this time, the upper stories were used as a hospital, and the basement and some of the accessory buildings were occupied by a government bakery. Although I had often admired the Seminary from a distance, I had no time or means to obtain reliable information concerning it.

One day of relief from routine was furnished by the grand review, and the army returned to its old ways.

November 14 – Found us on picket at Rose Hill. My post was on the beautiful lawn in front of the mansion on the hill. From this elevated position, the Dogue Creek can be seen winding through the valley for miles, fringed here and there by clumps of willows, and the hills that guard the valley on either side, clothed in their autumn verdure, complete a picture of rare beauty. With all the imposing grandeur that outwardly surrounds this fine mansion, there is sorrow within today. With the visible evidence of war at their door to prevent their egress or the ingress of kindly assistance, death has entered and stricken one of the household. The mother with only her young children in the family carriage, following the light wagon containing her dead, with no other attendants than the two Negro drivers, presents a strong contrast to the elaborate obsequies that would have taken place in time of prosperity. The utter loneliness of this family in their grief, without priestly council or the consoling sympathy of friends, brought tears to the eyes of more than one soldier though the natural protectors of these people were in rebel ranks.

The fine weather that had been with us so long was broken by one of those powerful rainstorms peculiar to Virginia. The little bay between our camp and Alexandria was raised so high that the bridge that spanned it was under water and communication by that route was suspended.

December 6 – A man with a two-horse team, with a Negro boy and two hogs in his wagon in attempting to cross the bridge into Alexandria was carried into the bay by the current. Our men tried hard to save them but could render no assistance. The man swam ashore and saved himself, but his property consisting of Negro boy, horses, hogs, and wagon were all lost.

The weary days dragged slowly away, and the soldiers growled at the everlasting round of duties without a chance to see the enemy. Once in a while, there was a reconnaissance. We privates could never see what a reconnaissance was for. A large body of troops were marched a long way out from camp, lay around two or three hours and then returned to camp without having done anything that we could see.

Every northern paper that reached us was full of clamor that the army should move and strike one blow before the good weather was past. Most of the soldiers felt the same way themselves, though faith in the patriotism and the ability of McClellan amounted almost to worship. Now the winter was approaching, and the rain was making Virginia one vast mud hole through which it was almost impossible to move artillery or supplies.

December 15 – The camp of our regiment was moved to a wide ravine, descending to and opening on Washington Valley, a part of George Washington's Hunting Creek estate. We were informed this would be our winter camp, and that we might use the timber adjacent to build such winter quarters as we could for our protection and comfort.

December 17 – The picket guard was formed by details of men from all the regiments in our brigade under the command of a captain from the 3rd Michigan. As there was no commissioned officer detailed from our regiment, it became my duty to report with the men from my regiment and in doing so, I found them on the left of the detachment. The captain, who was an entire stranger to me, directed me to take my position on the right. When I prepared to move my men to the right also, he good-naturedly ordered that they remain where they were, saying it was me he wanted on the right. We made the long march to the outpost with scarcely a word passing between us. Upon our arrival, he began to station the men by taking them off the left. When I requested to go with my own men, he laughingly told me he had special service and would see to it that I had good men. Having disposed of all the men but six, he conducted us two miles to the front where on an elevation, four roads, all leading into the rebel camp, spread out like the rays of a fan.

A small patch of timber occupied the angle formed by the junction of the road on which we had advanced, with these roads and in this we were posted with instructions to let no one pass or repass, and to give notice of the approach of an enemy. My remonstrance that he had placed me in a very responsible position with a small force of men that I knew nothing about was met with the easy remark, "You'll be all right."

Beyond the timber was the house belonging to the estate, and we had not long been on the ground before a little Negro brought a request for the officer-in-command to call at the house. Visiting the house, the lady told me plainly that she was a rebel and that her husband was a rebel officer and a prisoner of war at Fort Lafayette. She requested me to place a guard at the house to protect her and her property. Assuring her that my men would not harm her or hers, I told her that was all I could do at present. Before night closed in, the rain came down in torrents, and continued all night. In our isolated position, we dared not build fires, so we leaned our muskets and ourselves against the trees and wore the long night away.

December 18 – When day at last dawned, we found that every musket was full to the muzzle of water. We made all haste to draw the charges and wipe the barrels dry. Loading, we fired one volley, which was to be the signal that we needed help. Soon a cavalryman came dashing up from the reserve to learn the cause. I directed him to notify the captain that I absolutely refused to stay another night so far from assistance in such an exposed position with so few men. In the course of an hour, the captain arrived with forty men and jocosely inquired if I would be afraid with this addition to my command. When I told him in the same vein that with these I could make a creditable retreat, he said, "You'll do. I'll leave the men. Make yourself comfortable but be vigilant."

Before night, we were surprised to see General Phil Kearney riding toward us from the rear. Approaching and returning my salute, he inquired who was in command. Being informed, he seemed surprised and said, "Sergeant, do you know that you are in a very risky position?"

Receiving assurance that I realized the danger, and ascertaining all the particulars, he cautioned me to be watchful, and said, "We'll risk you tonight, and tomorrow you will move back to the line or the line will move up to you."

With the prospect before us of remaining overnight in our present position, we turned our attention to providing for the inner man. Selecting a couple of men that I had observed to be equal to most any emergency, I requested them to procure some potatoes and cabbages and a kettle. They soon returned with a good supply of the vegetables and a three-pail iron kettle. Collecting from each man his ration of bacon and as many hard tack as was needed, we soon had a wonderful mixture boiling over a good fire. When our stew was sufficiently cooked, each man helped himself directly from the kettle. This chowder was kept hot all night, and it was amusing to see a soldier rise from his restless couch at the foot of a tree, go to the kettle, gorge himself and retire, his place being taken by another and so on through the long night.

December 19 – When the hour arrived to relieve the picket, we beheld the whole line advancing. The line had moved up to us, and we returned to camp. The building of winter quarters engaged us for the next week. The men formed themselves into messes, and each little squad cut logs and laid them into a house to suit their own notions. One side was raised a log higher than the other to give a slant to the roof. Long slabs were split from chestnut trees and laid on top, then earth was shoveled onto the slabs and sodded to form the roof. The huts of the enlisted men all stood in a line on the other side. Some built stone fireplaces with stick chimneys. Other chimneys were topped out with headless barrels. The interior arrangements varied according to the taste, convenience, and ability of the builders.

The winter house of the eight members of my own mess was a fair type of all the others for interior furnishings. Our inside space was eight by thirteen. Setting a post in the center that had two crotches, one a foot from the ground and one four feet from the ground, we laid cross pieces from the crotches to the logs at the side, then we laid poles close together from the ends of the room to the cross pieces. On the poles, we spread rubber blankets. Then we went down by the Hunting Creek and cut long dry grass to make beds on the blankets. Over this we spread rubber blankets, and we had four passable bunks. As two men slept in each bunk, we had two woolen blankets for covering. Wooden stools, seats, and legs split from chestnut logs, a rough shelf for writing desk in one corner, and a sibley stove in the other completed the outfit. The sibley stove is made of sheet iron and is conical in shape. It is intended for use with the sibley tent which as an iron tripod at the bottom of the center pole.

Regular drills were suspended on account of the mud, and we dragged out an aimless existence in this little den, with nothing to speed the dull hours but cards, an occasional newspaper, a day on guard, forty-eight hours on the distant picket line once a fortnight, and sometimes a letter from home.

December 29 – Glad to escape from camp on any pretext, we found ourselves on the picket line, which had been pushed out to Pohick Creek. Pohick Church stands on the north bank of the creek seven miles from Mount Vernon. Washington drew the plan and was instrumental in locating and building this church. He was vestryman in this, as well as Christ's Church at Alexandria, and usually attended Pohick Church during his early-married life. Lord Fairfax, George Mason, and others of the noblest aristocracy of Virginia attended divine worship here.

The structure is square with hipped roof and without spire or belfry. Outwardly, its double row of windows gives it the appearance of being two stories high. Within what a change has taken place since its pews were filled with the greatest in the land. What a sight to meet within a church. The pews have all disappeared. Some battered remains of the once elegant pulpit stands at one side. The floor indicates that horses have been stabled here. The white walls are marred all round with rude vulgar scrawls done in charcoal. That the church has been a rebel cavalry camp, several rebels have left the assurance by writing their names and that of the command to which they belong with charcoal on the walls.

What would have been the holiday season at home had come and gone with us like any other time of year. There were no feasts or frolics. These days passed in with the unvarying round without even being mentioned as they passed. There was no Christmas or New Year for a soldier.

January 1, 1862 – The year began and continued in camp with no public events worthy of notice. There was little of interest to infuse life into the soldiers, of a private nature. There were some deaths among the privates, and these were buried according to the military ritual.

Squads of men were furnished with ammunition and permitted to practice target shooting under the supervision of a sergeant. The ground selected for the practice was on the flat along Hunting Creek. The target was placed on a gum tree four feet through and two hundred, four hundred and eight hundred yards were marked off to give a variety of ranges. Some that thought themselves good shots were surprised that they could not hit the body of the gum tree offhand at eight hundred yards. This practice was of great value afterwards, as the experience enabled us to determine how near an enemy must be before our shots would be effective.

Officers and non-commissioned officers during our inaction became slack in the attention to minor details of duty. I was prompted to closer attention to my duty in a way that I never forgot while on camp guard one disagreeable night. We had a large number of men on guard and the nature of the ground made a great deal of climbing up hill and going down in changing the relief. My relief came on at midnight, and when the roll was called, one man named Westbrook was absent. Any amount of calling and searching did not discover the man. I determined to post what men I had and fill his place afterward. Having relieved the entire guard but one posted at Colonel Poe's quarters, I told him to be patient, and I would relieve him at once. I returned to the guardhouse and found the missing Westbrook standing composedly by the fire. As I was very tired by the long tramp through the mud and cold, I administered a sharp rebuke to Westbrook and ordered him to go alone and relieve the man at the colonel's. He went as directed, but the sharp ears of the Colonel detected the absence of a sergeant when the sentinel was relieved, and summoned the officer of the guard and directed him to send me to my quarters under arrest. Next morning I was summoned before the Colonel who demanded by excuse for sending a sentinel to his post alone. When I told him I had no excuse for I knew better, but that it had been the practice for some time, he said to me, "I am aware of the fact and intend to break up the practice. I did not arrest you because it was you, but you were the first caught, and I shall trip up some of your superior officers soon. You are released and will return to duty." The Colonel kept his word and broke up this and other like practices.

About midwinter, our first installment of recruits arrived. I directed one fresh from Canada that was assigned to my squad to the cook shanty to get his ration. He soon returned holding his dozen hard tack between his thumb and fingers, and inquired if these were not all he required for supper, if he could have any more. Being told that was a day's ration, and that he could get no more for twenty-four hours, he protested he should starve. A mischievous soldier just then came up and asked him if rascally cook had given him his ration of butter and pie. It was very amusing to see the energy with which he returned to the cook and to hear him stoutly maintain his demand for the luxuries he believed the cook to be withholding from him.

The winter dragged slowly away with scarcely an incident to relieve the grinding of the military mill that turned out the same grist day after day. The men repeated the same old jokes and told the same old stories until they were very stale. The situation of the army and the probability of an early spring campaign were discussed until there was no more to be said. False rumors were invented and circulated until no one would believe anything he heard. The whole army was heartily tired of this idleness and was willing to take the chances of active warfare.

CHAPTER 5 – JANUARY/FEBRUARY 1862

The column of months drifted slowly by and still we lay snug in camp. Virginia was covered with deep soft mud through which it was next to impossible to transport supplies or move an army.

The newspapers were full of criticism on the Commander in Chief and urgent appeals for a forward movement. The disgust of the soldiers was about equally divided between the newspapers, generals, and their own enforced idleness. Judging from the tone of the press, most any country paper had a better general on its staff than was in the army. At the same time, soldiers serving in Virginia knew by experience the difficulty of marching through bottomless mire and the absolute necessity of following the army with supplies of food, forage, and ammunition.

March 8 – The first indication of active operations was the publishing of an order on this date, dividing the army into corps. This seemed to perfect the machinery of a grand army by adding a wheel to connect the more remote parts of the machine to the grand central power of the Commander in Chief.

The 2nd Michigan Regiment was assigned to the 1st Brigade, 1st Division, 3rd Corps. The corps was commanded by General **[Samuel B.]** Heintzelman, the division by General Phil Kearney, and the brigade by General J.B. Richardson. **[This is most likely another transcription error. By all accounts this would be Israel B. Richardson.]**

[The operational organizations of the Civil War were, largest to smallest: army, corps, division, brigade, regiment.]

March 13 – At last, a general order came to break camp and report to Alexandria to embark on transports. We were going somewhere and rumor assigned various places as our probable destination. We were rejoiced to go anywhere to break the monotony of our long idleness. Leaving our warm huts, we moved three miles toward Alexandria and camped on the bluffs overlooking the town. Our old sibley tents that had lain idle through the winter were found to be quite rotten, and we pitched them with many doubts of their ability to stand a strong wind. During the night, a disagreeable cold rain set in, and it was not long before the tents had absorbed sufficient weight of water to collapse them, leaving us out in the wet and chilly night. Making use of the soaked canvas for covering, we shivered the time away until morning.

March 14 – The dawn of day revealed Richardson's brigade, thoroughly drenched and chilled upon the bare, bleak bluff, without shelter from the pitiless rain. In this emergency, poncho tents were given us, and we were informed that this would be the army tent for the future, and that each man would carry half of one in his knapsack. That the idea filled us with dubious reflections will be made apparent by a brief description.

The half of a tent is merely a piece of ducking two yards square, with buttons on one edge and buttonholes on the edge at right angles to the buttons. The other two edges have eyelet holes with loops of marlin in them to drive stakes through. To put up the tent, crotches and a pole must be cut from the nearest timber. The pole being upon the crotches, two men button their halves together and stretch them over the pole. Spread the bottom as wide as it will reach and drive stakes through the loops. This furnishes a shelter open at both ends into which the men creep on their hands and knees. This is the protection against the dew and rain when it comes straight down, but does not keep the rain off the head or feet if the wind blows. Besides poncho, this tent is often called shelter tent, pup tent, and dog kennel. In the present instance, there was no timber near us, and the shelter tents were not available. Richardson asked permission of General Heintzelman to take his brigade back to their winter quarters as the transports had not arrived, promising to have them on hand at three hours' notice. Heintzelman's reply was philosophic, and helped us to reconcile many a sad plight. His harelip merely uttered the sentence, "It is incidental to the war, and the men have got to stand it."

Richardson returned and personally informed the men of the general's decision, and with his nasal twang somewhat intensified, said, "Boys, take care of yourselves." Availing themselves of this permission, the men were soon scattered all over the country in barns and sheds or wherever they could get out of the rain. For myself, I found shelter in an empty car shop in Alexandria and gathered there as many of our company as I could find, and notified the captain of our whereabouts.

The inimitable Westbrook, with two comrades, went to a farmhouse and ordered dinner, telling the lady of the house, "We want to pay you well for it." A fine dinner of chicken, warm biscuit, potatoes, and gravy was duly served and eaten. Then Westbrook said to her, "I allow you expect some pay for this." She reckoned that a dollar and a half would be the right amount to pay. Westbrook replied, "I told you we wanted to pay you well, and we do, there is nary a red in the party." Then they departed and left her to reflect.

March 15 – Our captain notified us to join the regiment as it passed through the city to the wharf. Vessels had arrived to transport us to our future field. Troops were embarking, and our regiment was soon on the Commodore Vanderbilt, which dropped off into the stream and anchored as soon as it was loaded. The embarkation continued until all the twenty steamers present were loaded to their full capacity.

March 16 – Day broke bright and beautiful, the clouds had rolled away, and the sun shone as it only can after a rain. The anchors were got up soon after sunrise and the transports, with orders to keep in line, were soon steaming down between the beautiful banks of the Potomac, a grand pageant. The decks were crowded with soldiers in brilliant uniforms, flags were fluttering in the breeze, the sun glinted from thousands of burnished arms, while numerous bands filled the air with music. The breath of spring was upon the earth, and the soldiers' hearts were glad—glad to be relieved from a tedious winter camp, glad to see a movement toward the great object for which we had taken up arms. Going we knew not where, and knowing that our course led us toward the rebel batteries that had lined the Potomac all winter, still we were glad.

About six miles below Alexandria, on the Maryland shore, stands Fort Washington. It is a stone fort, pierced for two tiers of guns with a third tier on the parapet. As we passed, the garrison gave us three hearty cheers, and the cannon boomed a salute of ten guns as a Godspeed on our mission. The tremendous answering that rolled back to them across the water was the outpouring of the pent-up energy and patriotism that filled us to overflowing. A little farther down, we passed that lovely spot that is sacred to all Americans—Mount Vernon, the home of Washington.

The first of the rebel batteries that we approached had already been taken by some of our forces that had moved by land, and the stars and stripes were run up as we came in sight. At nightfall, the fleet anchored; during the night, three gunboats joined us.

March 17 – Our course was resumed at daybreak under the convoy of gunboats. At Aquia Creek, there is quite a good harbor. The rebels had occupied the place, and here we expected to have trouble. As we drew near, a large cloud of smoke was seen ascending, and some cavalry was discovered on a hill watching our movements. A shot or two from the gunboats dispersed them, and a near inspection of the landing showed that they had evacuated, burning their supplies and warehouses. The remainder of the day and night was passed at anchor in the mouth of the creek. Many of the boys took this opportunity to take a swim, diving from the decks of the steamers. Soundings showed the water forty feet deep, and one man performed the, to me, wonderful feat of bringing up mud from the bottom.

March 18 – We left Aquia Creek and were soon on the broad bosom of the Chesapeake Bay. Brought up inland, as most of us had been, our twenty transports riding upon the wide expanse of blue saltwater, with many other vessels under full sail in the distance, was a great novelty, and the scene was enjoyed to the full extent. Late in the day, Cape Charles and Cape Henry were pointed out and full-rigged ships were seen standing into the bay from the Atlantic Ocean. A little later, we sighted the low tapering neck of land called Old Point Comfort. The channel between this point and Willoughby Point is the entrance to Hampton Roads, and the Rip Raps stand in the middle of the channel. Near the end of Old Point Comfort stands Fortress Monroe, a large and formidable stone fort. The Rip Raps, a rocky island, is entirely covered to the water's edge by Fort Calhoun. The fort, which is unfinished, had been building by convict labor for forty years.

Vessels entering the Roads Pass between these two forts with their heavy guns and the Union gun, carrying a round shot of four hundred and twenty pounds weight, were mounted upon the beach near the water's edge on Old Point Comfort.

Hampton Roads is a sheet of water about ten miles square. The entrance is at the northeastern corner. At the southeastern is Norfolk near the mouth of Elizabeth River. The Nansemond River empties in at the southwest and the James at the northwest. Within this landlocked harbor, hundreds of steam and sail vessels of every shape and occupation and of various nationalities rode at anchor as we sailed in and cast anchor in front of Fortress Monroe.

Men-of-war of the Russian, English, French, and American nations were about us, but more curious to us than all was the little *Monitor* **[USS Monitor was an iron-hulled steamship and the first ironclad warship commissioned by the U.S. Navy during the Civil War. Its success in warfare made wooden ships obsolete in wars.]** that had so lately vanquished the terrible *Merrimac* **[later known as the *USS Virginia*]**. It lay close to us that we could talk with the men on board. Her appearance was quite novel. All there was visible was what looked like the deck of a vessel floating upon the water with a flagstaff at one end, a small box to protect the steering wheel or pilot at the other end, an iron turret a little forward of the center. We were full of admiration and gratitude for the little "Cheese Box on a Raft," but could not understand how she did it.

March 19 – The troops disembarked and camped on Hampton Creek near Hampton Female Seminary and ex-president Tyler's summer residence. Neither building was enough out of the usual to attract special attention.

The forts at the entrance of the Roads, Suval's Point, Newport News Point, and the mouth of the Elizabeth River were in plain view from our camp. Hampton Creek broadens at its mouth and forms a little shallow bay, the waters of which are made just brackish enough by the tide to successfully raise oysters. Observing a Negro out in a boat raking the bottom with two rakes joined like a pair of shears, I waited for him to come ashore and discovered that he had oysters. Asking him for the loan of his rake and boat, he replied, "I'se got no power to hinder yer takin' 'em. Der dey is." I was quickly over the bed and soon had all I could carry away. Coming ashore, another soldier met me and borrowed the boat for the same purpose. I had got far enough to safety, but was yet near enough for observation, when General **[John E.]** Wool came dashing up on a powerful white charger.

He was dressed in full uniform with the exception of a glazed cap from beneath which his gray hairs were peeping. With majestic sternness, he hailed the soldier in the boat and demanded, "Don't you know those oysters are private property. They were planted there for private use. You have no more right to take them than you have to steal their garden truck. Come ashore this instant." Down went the rake again and up came more bivalves, while the wind wafted across the water, "Go to Shore." Wool's long experience with regulars led him to expect implicit and instant obedience, as well as respect. His frenzied rage was fine to see when he heard the reply of this lawless volunteer. Promising to have him in irons in fifteen minutes, he rode furiously to the fort for a guard. I lingered to see the outcome, but when the guard arrived, the boat and rake lay securely on the beach, and the soldier and oysters were gone. The next two days it rained almost incessantly and a raw cold wind blew from off the water. Our poncho tents afforded but little protection.

March 21 – The bugle sounded "Strike Tents." Taking the road up Hampton Creek a mile or so, we crossed the bridge and passed through the ruins of Hampton. This was the oldest town and had the oldest church in Virginia. The rebel cavalry general, Wade Hampton, occupied this place through the winter with the Hampton Legion. Upon the approach of the "Yankees," he fired the town and retreated. The town was entirely destroyed, but the greatest regret felt for the loss of the old church, said to have been built by John Smith of brick brought from England.

Leaving the blackened ruins of this little hamlet behind, we soon pitched our tents near Back River about five miles form Newport News Point. In this camp, we passed the next two weeks. The rain fell almost continually and the ground became a sea of soft mud through which it was difficult to move. Troops were arriving and going into camp around us every day. The enemy was in possession of the other side of the river and picket duty came around often. A night on picket in the rain was not much relieved by a day under our pup tents. We were not yet used to accommodating ourselves to such close quarters, but it was discovered that a third half tent, turned corner-wise would button on to the others and close up the end. This and other discoveries afterward enabled us to make ourselves comparatively comfortable.

I had often noticed in Virginia a small thicket in the midst of cultivated ground. In an intermission between showers one day, I explored one of these and found it to be, as I did others after, a family burying ground. The tombstones were common sandstone and lay flat upon the ground. Rudely chiseled upon their rough surface was the name, date of birth, date of death, and sometimes a brief epitaph. The date of death on one was 1711. I saw others ranging along to 1790. It seems disgraceful that these old burial places should be neglected to grow up on tangled thickets and plowed around like any difficult obstruction.

While our regiment was on picket along the river one day, men were discovered lurking in the woods on the opposite bank. A peremptory order enforced with loaded muskets brought them over the river. There were seven of them, and they proved to be Negro slaves escaped from Yorktown. They were in a starving condition, and one of them died the same night. Twenty-four of them had escaped from Yorktown, an easy day's march from us, four weeks before. They had concealed themselves by day in ditches and wandered by night in search of the "Yankees." Seventeen had died by the way from hunger and exposure, and one died just after reaching the goal of their desire. They gave important information concerning the enemy's fortifications, position, and strength. The survivors were employed as servants by our officers. This incident shows that slaves will risk their lives for freedom.

April 4 – The long expected forward movement was begun at last. The clouds that had poured rain upon us so long and continuously had rolled away, and a fiery sun shone down upon us with all the fierceness of full summer. We crossed Back River and took the road up the peninsula between the York and James rivers.

The roads were heavy with mud, and the soldiers were fain to relieve the tiresome march by reducing the weight of baggage carried upon their backs. Believing summer had come to stay, they began to throw away their overcoats, and some their blankets and soon the roadside was literally covered with these castoff articles. The route step of vigorous, fresh troops soon brought us through a wooded country to Little Bethel. We were some surprised to see nothing but a small church, for General **[Benjamin F.]** Butler's movements of last June had given the Bethels a place in history.

[It's noted that General Butler was one of the first to employ observation balloons during the Civil War, but his reputation as a terrible tactician was proven on June 10, 1861, at Big Bethel, Virginia, which was the first land battle of the Civil War. Despite outnumbering the Confederates, his men fired on each other, which

accounted for one-third of the Union casualties in the defeat.]

Late in the afternoon, we came upon Big Bethel. Here as before, there was a church though a little larger than the other Bethel. Some deserted rebel earthworks stretched away into the woods on each side of the road, but as there were no rebels about there was little of interest to us. Two miles beyond Great Bethel, we bivouacked for the night, sleeping beside our arms, for it was known that we were very close to the enemy.

April 5 – At an early hour, the column was on the move. The men, in high spirits, pushed eagerly forward, the dominant thought being that every step brought us nearer the enemy. The road lay mostly through heavy timber. In the after part of the day, the country became low and swampy and the roads soft, the mud coming often to our knees. The distant booming of artillery freshened our lagging zeal as we came upon higher ground, but a brief delay, during which the sounds of strife at the front became more animated, raised hopes that once more we were to have some practical experience in warfare. As the firing died away, we moved on and soon learned that the enemy had made a feeble attempt to check our advance at Worms Mills.

Late in the day, the occasional discharge of cannon sounded nearer, and as we emerged from the woods onto an open plain, we could see the smoke of the guns rise from a dim line of earthworks on the opposite side. That we were within range of their artillery was evident from the shots that now and then screamed over our heads and clipped the limbs from the trees in our rear. Just at twilight, our brigade was moved to the left and camped between the woods and a pool of water. Beside the pool were Professor Lowe and his balloon. As we approached, he was being pulled down from his lofty place of observation, a thousand feet high. This was one of McClellan's favorite ways of gaining information of the enemy's movements.

As we had experienced a long and hard march, the cooks were ordered to serve coffee with dispatch. Being in a strange locality close to the enemy and already dark, the cooks took water for the coffee from the pool. When the coffee was served, each man, as he took some in his mouth, quickly spit it out again. Everyone declared it bit his mouth so they could not swallow it. Investigation showed that they had taken the water from the same spot where the refuse vitriol used in manufacturing gas for the balloon had been emptied. A storm of indignation was raised against the cooks and some curses were hurled at the balloon, but I doubt any of us could have done better under the circumstances.

We were now before Yorktown and on the border of the plain where Cornwallis surrendered his forces in Revolutionary times. Yorktown is a walled town of the old style to which the rebels added a long wing of earthworks that entirely covered the available route up the peninsula.

After eating the remnant of food in our haversacks, we lay down with the thought that on the morrow we should surely assault the enemy and severely punish him.

April 6 – With the dawning of a new day, the encampment was astir, anxious to get a glimpse of the surroundings by daylight. The first object that met my gaze as I crawled from my shelter tent was the forms of General [Hiram G.] Berry and two staff officers stretched upon the bare ground under a tree, without other covering or shelter than its spreading branches while their horses gnawed at neighboring saplings. The baggage train that should have brought the general's tent and equipage and our rations, had been delayed by the bad roads. The prospect for something to eat was dubious. The majority of the men had eaten the last from their haversacks the night before. Roused by the bustle of the camp, the general and his aides arose and kindled a small fire and put something to cook in a quart cup. Three of us who messed together pooled our rations and found we had six hardtack for the three. **[Hardtack is a type of cracker and has been the staple of militaries for centuries. It consists of flour and water, and perhaps sugar and salt, if available.]** We voted to give the general and each of his aides one apiece and have one each for our own breakfast. Upon tendering them to the general with the information that it was half we had, he accepted on condition that we take half of the rice they were cooking which was all they had. It was a great consolation to us that for once the general and his soldiers fared alike.

Notwithstanding our eager expectation, the day passed without any action. The next day and the next passed in inactivity and no rations.

[Elsewhere in the war, General Ulysses S. Grant's troops are surprised by a Confederate attack at Shiloh on the Tennessee River. More than 23,000 men are killed or wounded.]

April 8 – The troops were moved back into the woods and picks and shovels came to the front. Our spirits were dampened by the prospect of more shoveling. For many days, camp life was broken only by the regular rotation of picket duty and an occasional turn at shoveling in trenches, which were to form a regular approach by means of parallels to the enemy's works.

The camp of our brigade in the woods lay between two swamp holes, the waters of which we were compelled to use for washing, cooking, and drinking purposes. Very soon around the entire edges of the ponds was a wide circle of dirty soapsuds where the men had washed their clothes and their persons. Water to drink was procured by walking as far out as possible on a log and dipping up the filthy swamp water, and even there it tasted of soap. We knew of no better water, and if we had perfectly patrolled, that if a man strayed from his own camp, he was sure to go to the trenches to work instead of finding water. Drenching rains fell most of the time, rations were scarce, and the miasma of the swamps sickened the men so that our ranks were depleted faster by disease than they would have been by assault upon the works at Yorktown. At intervals through all these days, the rebels sent shot from their heavy guns over us and around, but nobody was ever hurt by them. Working parties pushed the fortifications day and night under the direction of experienced engineers. By day, the work went on behind the woods and in other places out of rebel observation and by night parallels were dug across the open spaces by the aid of white lines. Each morning, the enemy awoke to find some new piece of Yankee impudence right under their noses.

One dark and rainy day, I was ordered to take out a working party of forty men. The engineer conducted us to a narrow strip of woods that projected into the plain on the same level and in full view of the walls of Yorktown. On the edge of the wood, next to the enemy, was a fine growth of underbrush. Through this as close as possible to the open ground, the engineer ran his line, his own men cutting off the fine brush and standing it up to thicken the screen that concealed us. It was our work to follow and cut out the trench, one spading deep so that a larger force could throw out the remainder at night without the aid of lights. With a wholesome respect for the frowning guns before us, we did not draw a single shot from the rebel batteries.

As fast as works were completed, men were detailed to man them for twenty-four hours at a time. It was my fortune to be in the same trench several times during the thirty days siege of Yorktown, and I witnessed two or three amusing incidents. It was the custom of the rebels to fire a shot once a half hour over this work, and sometimes it hit the work and at others, it clipped the treetops. Just behind us was a low spot where grew some lofty elms. Through these, one day came a straggling soldier leisurely surveying the surroundings. He paused in a comfortable attitude under one of the elms and seemed to be enjoying himself very much when a shot from the enemy cut a limb as big as his body directly over his head. He did not wait for the limb to fall, but started on the jump for camp and had not slackened his speed nor looked behind when he entered the wood on the opposite side of the plain. At another time, General Phil Kearney and three other generals rode up to this part of the line and stopped to gaze through the thin growth of timber at Yorktown. As they looked, a puff of smoke shot up from the rebel works. Quick as a flash of light, the four generals were off their horses and flat upon the ground. When the deadly missile had passed, they arose and remounted. Kearney remarked with a pleased smile, "I never before saw generals so quickly reduced to the ranks."

I came on duty once with a captain of the 3rd Michigan. After lookouts had been posted and everything was arranged for the night, the captain and myself wrapped our blankets around us and lay down on the shelf of the trench, feet together, while the other men were disposed conveniently near. In the night, a big shell buried in itself in the breastwork near the point where our feet met, and burst, sending a great amount of dirt into the air. As the earth came down, it buried both under a heavy load. With some exertion, we both released ourselves and met in the darkness, he being sure it had torn him to pieces. Investigation showed that no one was hurt, and we returned to our slumbers to be disturbed no more that night.

Time brings all things to an end and at last the earthworks were completed, the cannon and mortars mounted, and it was currently rumored that the bombardment would take place in a day or two. Our sharpshooters were so close to the walls of Yorktown that they could pick off the gunners whenever they attempted to serve their guns, and their keen eyes were always strained to detect an animated mark for their aim.

May 4 – When daylight illumined the world, no gaze was sharp enough to see anything astir along the rebels' works. Some of the bravest of the advance pickets crept forward to reconnoiter. Meeting with no opposition and emboldened by the solemn stillness, they took a peep through the embrasures to see what the "Johnnies" were up to. No enemy was there. They had evacuated, leaving some tents standing and other camp equipage in their haste to get beyond the range of our guns. Also, the heavy guns on their works were left shotted, ready to fire. The temptation was too great for one of the adventurous men who first entered the works, and he touched off a gun that was pointed directly over our camp. The whole army was aroused from its slumbers by the thunder of the heavy gun, and the shrieking of the ponderous shell as it cut its way through the treetops. This was the last gun from Yorktown, though fired by Union hands, and did no harm. News of the evacuation flew like wild fire and soon the whole army was in pursuit of the retreating rebels.

They left behind some relics of barbarism in the form of torpedoes and bombs with percussion caps, easily exploded when stepped on. These were planted where soldiers were liable to stray and several were killed by them before they were all discovered and carefully removed. Our brigade went into camp four miles beyond Yorktown before dark.

CHAPTER 6 – MAY 5, 1862

With the dawning of day came a pouring rain that continued steadily all day and far into the next night. The soldiers were early roused and prepared their breakfast and ate it amid the deluge.

In brief time, they were marching toward Richmond through the deep clay mud that was momentarily growing deeper as each successive command stamped it into paste with the falling water.

The rebel army had retreated over these same wet roads with their artillery and wagon trains and the advance of our own cavalry and artillery had followed the wheels going down to the hub and the horses to their knees. Now we were literally wallowing knee deep, often deeper, through the bubbling mire, while the splashing and heaving up of the mud plastered the men to the waistbands.

About eight o'clock, there broke upon our ears the sound of heavy cannonading mingled with the roll of musketry. As we floundered on at the rear of our division, the din of battle grew more distinct and seemed to increase in extent. Often we were delayed by the troops in advance and chafed at the obstructions that kept us from the scene of conflict. Toiling on wet and weary, we were fearful as the day dragged away, that night would come, and we have no part in the fortunes of the day.

At length, a dispatch came from Heintzelman to press forward with all possible dispatch. Another and another appeal for speed did the lion-hearted old warrior send. The gallant General Berry could stand the delays no longer. Leading our brigade through the woods, we wallowed past two whole brigades and came upon Heintzelman on the edge of the battlefield at four o'clock. He was all alone and sat upon his horse swinging his saber and grinding his teeth. At our approach, he exclaimed, "For the love of God, men, move forward."

The fighting was going on in a slashing of many acres where the heavy timber had been felled so as to form a tangled obstruction at the junction of five roads. Out on the open plain beyond the fallen timber was a chain of forts extending across the peninsula. Fort Magruder, the largest and strongest, was so located that it commanded all of the roads. Two miles beyond in plain view was the town of Williamsburg.

In this slashing, **["Fighting Joe"]** Hooker's men had maintained an unequal contest all day under a galling fire of musketry from the rebel infantry, the iron hail of Fort Magruder and the pouring rain from heaven. Their ammunition was most gone and the men were worn out and exhausted. Disheartened, but still stubborn, they were slowly falling back when we arrived. General Phil Kearney, leaving the remainder of the brigade to be disposed of by Heintzelman, ordered the two companies D and G from the right of the 2nd Michigan detached for special service, to be led by himself. While the other regiments were being moved to the relief of Hooker's men, Kearney deployed our two companies on each side of the road leading straight toward Fort Magruder, and under his personal direction, we rapidly worked our way through the fallen timber toward one of our own guns that was stuck in the mud and which the rebels were trying to turn upon us. Tired, wet, and chilled as we were, there was no flinching. Guided by the alternate encouragements, cautions, and praises of the intrepid Kearney, we had soon passed through Hooker's lines, and the accurate aim of our deadly bullets forced the enemy to give ground. At each retrograde movement of the enemy, Kearney would shout, "They run. Up Michigan and charge 'em." If we were too near them, he would come and "Down Michigan. They'll hit you." The skillful maneuvering by Kearney of this handful of men recaptured the lost gun and pressed the enemy back to the edge of the slashing before night closed in and put a stop to the battle. (In his official report, General Phil Kearney says, *"Colonel Poe's 2nd Michigan held the key point of our position. Two of his companies led off with the success of the day."*)

With the lines firmly established, Kearney turned his attention to his surroundings and in so doing, an amusing incident occurred.

To the right of the road, the Mozart Regiment was fighting bravely enough, but their colonel was laying behind a large gum stump several rods in the rear, roaring out continually in his powerful bass voice, "Go in Mozarts." The rebels had caught up the catchwords and were tauntingly flinging them back. Kearney, guided by sound, hunted up the gallant Colonel and approaching him from the rear, he lifted him from the ground by a vigorous kick from his horseman's boots, at the same time howling at him, "Go in yourself, you blankety-blank coward. I have just been with a hundred Michigans where I could not go with ten thousand such men as you."

With darkness came fresh troops to relieve us, and we withdrew to the woods in the rear to rest as best we could through the cold rainy night, expecting to renew the combat in the morning. Without shelter, food, or fire, the long dark night dragged on with its ceaseless rain and the soldiers shivered it through in the mire of the swampy woods with little rest. (Union loss – killed 456, wounded 1,400; Confederate – 1,000 killed, wounded 166.)

[The Battle of Williamsburg, May 5, 1862, involved nearly 73,000 soldiers and resulted in the Confederate's withdrawal after General Kearney's division stabilized the Union's position.]

May 6 – At daylight, the rain ceased, and the clouds rolled away. The sun, bright and beautiful, sent its piercing beams through the trees glittering with raindrops. The sight of the sun was a joy to the soldiers' hearts, but the thought of the dead and wounded lying out there in the slashing was a gloomy background to all present pleasures.

The early morning scene among the trees was weird and supernatural. The warm sun made the dripping foliage to steam and its rays glinted sharply from the stacks of burnished muskets, while the soldiers with sober faces were moving quietly about their little fires, cooking their coffee, and the water laden leaves sent down a solemn drip through this scene of mist and flame and smoke and animate life, that seemed like mourners' tears for the fallen braves who were ready and waiting for soldiers' graves.

Before we had eaten breakfast, news came that the enemy had retreated. Ere long, we moved out of the dripping woods into the sunlight on the plain near Fort Magruder. The sun dried our clothes and warmed our chilled bodies and new life stimulated us to view our surroundings. Evidences of a hasty retreat were on every hand. Beside the road lay the knapsacks of a rebel regiment in perfect line as if the soldiers had dropped them off their backs at the last minute before hasty flight. Overhauling these knapsacks furnished occupation for a brief time, but the plunder found did not pay for the trouble.

While the main army was pushed forward in pursuit of the enemy, our brigade was detailed to bury the dead. This duty gave us an opportunity to see the horrors of war as exhibited by the field the day after the battle. No description can convey a just idea of the scene. It must be seen to be appreciated and sight alone can convey its ghastliness. The solemn stillness is in marked contrast to the deafening din and thunder of yesterday.

Groups here and there are searching for the bodies of friends among the slain that lie scattered promiscuously over the plain and among the fallen timber. Death has come to them in every conceivable attitude, and they remain in the same position, as they were when life ceased. The cramped body with distorted face, froth oozing from the mouth, eyes glaring and the surrounding turf torn and mangled indicate that life terminated in terrible agony. Others lie with a smile upon the lips as if to die in such a cause was a pleasure. Here is a headless trunk, there a body without limbs and yonder a body that a ruthless shot has nearly severed in twin. Under a tree, where friends placed him and composed his limbs for burial, there lies the body of a staff officer in a costly uniform with a bullet through his brain. A face with glassy eyes peers at us through the forks of a fallen tree. Over a log, we find that the face belongs to a dead soldier who has not closed his watchful eye, even in death. Astride a low broken fence, sits a soldier with his back against the stakes, his gun across his knees. A near approach reveals the fact that a swift bullet has pierced his brain and stopped him in his attempt to cross the fence. Along the rifle pits at the edge of the woods, there was a desperate struggle, and the mangled corpses are in heaps, ten or twelve deep. At the edge of the slashing, where the light artillery was pushed far to the front, horses and men lie dead together. Scattered in endless confusion among the bloodstained dead, are firearms and all the trappings of war. The earth is plowed up, and trees are shattered by the fury of shot and shell. But it was not for us to dwell on this sickening sight. We must harden our hearts and brace our nerves to bury the fallen out of sight in the earth. Deep and wide trenches were dug. These were filled nearly full of bodies, friends, and foes together. Then the earth was heaped upon them.

The arms and equipments were collected and the only signs of the conflict left on the field were the furrows plowed by the plunging shots and the splintered trees. The sun shone as brightly as it would if there were no graves to shine upon.

May 8 – I was ordered to report for duty with four men at Kearney's headquarters at William and Mary's College, Williamsburg. When we arrived, a staff officer informed me that we would not be needed until night and that we might make ourselves comfortable during the day under the trees in the yard.

The place was quite pretty and contained perhaps a thousand inhabitants. Most of the citizens had gone to Richmond for fear of falling into the hands of the Yankees. The college was the principal support of the town. The college buildings were not unlike those of the average college of the north. While we were enjoying a delightful rest on the shady lawn, the sound of loud and angry profanity at the front door of Kearney's quarters attracted our attention. A sergeant was flying down the walk with Kearney at his heels, kicking him at every jump and calling him all the vile names in the slang vocabulary. By a vigorous effort, the sergeant got beyond the reach of Kearney's boot by the time they reached the middle of the street, and then he turned upon the general with a torrent of obscene abuse. Kearney glared upon him for a single instant, then with a grim smile commanded, "Come here, Sir. Follow me." As he led the way directly into his quarters, our anxiety was intense to know what would come next. Drawing near to an open side window, we saw Kearney stalk up to a sideboard, fill a glass with liquor from a decanter, and hand it to the sergeant. He then said, "When you want to read, don't sit in my front door, but go out in the shade like those other soldiers. Go, Sir."

The sergeant swallowed the liquor, saluted, and withdrew, having learned a valuable lesson by hard experience.

At sunset, a staff officer conducted us to our station for the night. Our duty was simply to allow no one to approach a two-story gristmill that stood on a pretty little stream in the outskirts of the town. Our presence seemed to be all that was necessary, for we saw no one during the night.

May 9 – Early in the morning, we were relieved and joined our regiment, which was ready to move. Following the Williamsburg and Richmond Stage Road, we marched by easy stages during the next week through the little hamlets of Barhamsville, James City, a county seat with its slave pen, blacksmith shop, and two dwellings, and New Kent Court House, another county seat with the usual slave pen and perhaps a dozen dwellings, to the Pamunkey River.

May 16 – We camped at Cumberland Landing on the Pamunkey. The river is navigable for first class vessels several miles above this point, but it is very narrow. It is so narrow that vessels cannot turn around except in places where the river widens beyond its usual width. Two or three steamers were here from New York, unloading sutler stores. We indulged our appetites for luxuries as long as our money held out and then returned to army fare.

Not far from here is the famed White House where Washington left his faithful servant to hold his horse while he wooed and won Martha Custis. The house is large and white, but otherwise is not remarkable in appearance.

May 19 – We moved camp to Baltimore Crossroads where we lay idly listening to the occasional skirmishing at the front for a week.

May 25 – We broke camp and moved across the Chickahominy Creek at Bottom's Bridges and went on picket on Dr. Carter's plantation. Immediately in front of our line was a narrow, impassible swamp with timber beyond. The enemy's pickets were in the timber from which they kept up a rattling fusillade upon us all night. We could see nothing but their ordinary bullets and from the plunging sound of some of their shots, it was evident that some of them were firing from the top of the large elms on the other side.

We remained very quiet and queried some as to what we should do when day light should come and make us a fair mark for their aim. But as day began to break, their fire slackened and finally ceased. They were evidently as loath to be seen by daylight as we were. We remained here two days, but saw nor heard nothing more of the enemy.

May 27 – We camped beside the railroad two or three miles nearer

Richmond and perhaps three miles from Fair Oaks station. The stage road and the railroad run between a miry swamp on one hand and White Oak Creek on the other. At this point, White Oak Creek is a sluggish stream with no particular channel, that spreads itself over a brushy bog about forty rods wide and cannot be crossed except where fords have been built by laying in brush and hauling sand upon it.

Near us were two fords about a mile apart. The lower one was called Crittenden's Ford and the upper Jordan's Ford. To cross these fords, it was necessary to wade forty rods in water three or four feet deep. To protect our left flank from the enemy, these fords were constantly guarded.

May 29 – Companies D and G went on picket at Crittenden Ford.

The land along White Oak Creek is timbered with a heavy undergrowth and the brush in the creek itself is so high that it is impossible to see across. A small squad of men was stationed close by the ford to keep close watch while others just lay back with harness all on and arms in easy reach. Two days and nights, we remained here keenly alert, taking note of every snapping twig or rustling leaf. But nothing appeared to us to cause us harm or give us adventure.

May 31 – Orders came for us to report to General Kearney at Seven Pines Tavern. Without delay, we were on the move. Before we reached the stage road, one of those sudden storms peculiar to the south burst upon us without warning. The sky grew dark. Then quickly came sheets of livid flame, followed by deafening crashes of thunder. In another moment, sluices of water began to pour. Darkness became so intense that nothing could be seen except by the blinding, hissing, crackling flashes of lightning. The scene was one of terrific grandeur, but exposed to its fury as we were, it was not pleasant. Some gained the partial shelter of the trees. Others could not make head against the flood and were forced to stand and take it where they were. In half an hour, this cloudburst was a thing of the past. The only evidence that it had been was the distant detonation of thunder and the lake of muddy water in which we stood over our shoe tops.

As soon as the storm swept by, we marched away in pursuit of orders. Then there broke upon our ears rapid explosions of thunder that we knew too well were not from heaven, followed by an unsteady roll that we knew was not the reverberation of thunder along the clouds. To our experienced ears, it was the sound of deadly strife.

Then came fugitives from the front, saying that Casey's division which was in the advance had been surprised at Fair Oaks station and "All cut to pieces." As with increased pace and quickened pulse we pushed forward, the number of fugitives increased and all had the same cry. "We're all cut to pieces." To say that our little band felt no misgivings in the face of this wild rout would not be true. Thoughts of Bull Run forced themselves upon us, but when did the 2nd Michigan fail to report wherever they were ordered?

Straining toward the front, we met the lion-hearted, firm and true General Heintzelman at a point where the swamp and creek came close together within forty rods. This hair-lipped old general demanded, "What are these and where are you going?" Being told that we were two companies of the 2nd Michigan going to report to General Kearney, he ordered, "Deploy across this muck and stop these stragglers or kill them." Instantly, the movement was begun at double quick and in another moment, we were facing the mob of excited, terrified men, some hatless, from they knew not what, while the spent balls from the enemy was stimulating their speed.

To stay this tide was to us a harder task than to fight the enemy. They were our friends, and we did not want to hurt them. By the sounds from the front, we knew that our men who had not been stampeded were bravely holding the rebels in check. These men must be made to turn and help them. At first, it required rough treatment and some received wounds here that had escaped unscathed at the front, but when the tide was once stayed, a peremptory order to "Fall in," enforced by the point of the bayonet, backed by a loaded musket was obeyed without resistance. Each had his story to tell, to which we would not listen. Officers and men alike insisted that "We're all cut to pieces" and "I am the only man left of my regiment."

Officers and men resorted to various subterfuges and tricks to get past our line. Two men carrying their brave and esteemed captain, with both legs tied up with handkerchiefs, were stopped to examine the captain's wounds. When the bandages were removed, no wounds were to be found. Men with heads, bodies, legs, and arms tied up were detected in the cheat and put into the ranks. A colonel of a New York regiment with two men carrying him desired to push through. We sent the men to the ranks, but passed the colonel. He was dead-drunk. We dumped his carcass on the ground in the swamp as of no use. One by one, seven color bearers drifted back to us with their colors and the declaration that they alone had escaped with the colors, the others were "all cut to pieces." The phrase "cut to pieces" became a joke and many an officer in splendid uniform was asked to take off his clothes and show where he was cut. Some officers were indignant that their rank was not respected, and that private soldiers dared to prevent their passing, but a look into the muzzle of a loaded musket with a resolute eye behind it inclined them to waive their rights for this once. By stationing the various regimental colors in different parts of the field, and directing the men to assemble around their own colors, we rallied seven good-sized regiments of live men that were not "cut to pieces." We kept our line all night, part of the men sleeping at a time. Our duty had been a very unpleasant one, but we were assured that it was very important.

June 1 – Early in the morning we joined our regiment on the battlefield. Seven companies of the regiment were in the thickest of the fight and lost heavily in killed and wounded, and Colonel Poe had his horse shot under him. Richardson's division was already pushing the enemy, and long before noon, the lost ground was regained. This two-day battle was called by both names – Fair Oaks and Seven Pines, the fighting being done between a railroad station of the former name and a country tavern of the latter. (The aggregate loss to Union and Confederate – killed 3,690, wounded 7,524, prisoners 2,322.)

[Fair Oaks or Seven Pines, May 31-June 1, 1862 with the total killed, wounded, or captured now recorded as 13,736.]

Our entire regiment went on picket at Jordan's Ford of White Oak Creek. The surroundings of this ford are similar to that of Critttenden's. Both are on cross roads from the Williamsburg and Richmond Stage Road to Charles City Crossroad. This ford was watched with the utmost vigilance, for the country on the opposite side was open to the enemy. Should they effect a crossing here, the flank and rest of our army would be exposed to assault and such an event would be likely to bring disaster to our cause.

June 2 – General Kearney sent an order to Lieutenant Colonel Williams, who was in charge of the picket at Jordan's Ford, to send some trusty men across the ford to examine the Charles City Road where it crosses a small run, to ascertain if troops had passed on that road and, if any, probably how many.

Sergeant Boughton of Company G, who had gained some notoriety as a scout, was ordered to select four men of his company to go with him on this duty, and I was ordered to select a like number to go with me on the same errand. I was to act in conjunction with Boughton, or independently as my judgment dictated. The water in the ford was above our waists, obliging us to hold our equipments above our heads.

Coming up onto the high ground on the other side, a broad expanse of cleared land stretched away to the Charles City Road, a mile distant. With a plantation house midway of the opening on the road, we decided that Boughton's party should skirt the opening to the right, and I to the left, meeting at the house, each to come to the assistance of the other if firing should be heard.

Keeping under cover of the timber, guns as a "ready," keenly alive to all our surroundings, we progressed cautiously. An innocent hog that rushed with a snort from a thicket came near to losing his life, but without other incident, the road was reached and creek found.

There had been a slight rain the night before, and the damp sand showed the hoof marks of perhaps a dozen horses headed toward the rebel camp. Having accomplished our mission, and thinking the other party would be waiting, we went directly up the road to the house, entered the front gate, and passed around to the rear.

Sitting composedly upon the back porch, we came face to face with six lusty gentlemen dressed in grey. It was a surprise to meet enemies where friends were expected, but their pleasant "Good morning, gentlemen," and a return of the compliment with an observation on the excessive heat of the day, gave time for some lightning calculation. Our forces were about equal if they had no reserve force in the house. No great question would be decided by a fight between these small squads of men. The instant decision was peaceful separation, if possible. To our inquiry if there was any large body of Confederates near, they replied only some stray cavalry scouts. They said they were waiting for dinner. Would we set our guns against the trees in the yard and join them. We declined their polite invitation on the ground that the rest of our command was waiting for us down the ravine that crossed the field back of the house. Bidding them "Good day," we took our way toward the ford, the men in advance all the time pretending to talk with those in the rear, but in reality keeping strict watch of the men in grey. When we got into the ravine out of sight, we got to the ford without delay.

What was in the minds of the men at the house, of course, we did not know, but they did not seem to desire a doubtful conflict more than we.

At the ford, men were felling large trees across the road to obstruct the crossing of cavalry. They told us that Boughton's party had recrossed half an hour before. They had seen the men on the porch and did not go to the house. Recrossing the creek, we reported what we had seen and joined our company.

June 3 – Released from picket duty, we camped on the edge of Fair Oaks **[Seven Pines]** battlefield. The work of burying the dead was being pushed with vigor.

Orders against straggling were very strict and anyone caught away from his own camp was put to grave digging or grave filling. On this account, those not actively engaged in the work of burial or caring for the wounded saw very little of the horrors of the field. A comrade and myself did venture as far as a field hospital, not far from us, and witnessed as much as we cared to look upon. The hospital proper was nothing but a large tent fly or cover under which stood dissecting tables around which the surgeons were busy amputating limbs and dressing the more difficult wounds. On the ground under the trees were long rows of wounded men, some waiting their turn on the dissecting tables, others being cared for where they lay by the nurses and hospital attaches. The sight was terrible. On every hand was bloodstained misery. Those pallid, pain racked, uncomplaining countenances can never be forgotten.

June 4 – General J.B. Richardson, accompanied by General **[William Henry]** French, rode into our camp in the familiar way he used to when he was colonel of our regiment, to tell the boys he said, "How we've got things fixed at the front."

Richardson carried in his shoulder a ball received in the Mexican War, which caused him to continually hitch his shoulder. General French, in the same service, received a wound in the cheek, which caused the muscles of his face to twitch severely whenever he spoke. The boys gathered around to hear what Richardson had to tell and every point he explained, he would turn to French and ask, "Ain't that so, General." The peculiar jerking of Richardson's shoulder and the comical twitching of French's face afforded any amount of amusement for the boys, but they had too much respect for the generals to let them see what amused them. All waited until Richardson said a good thing, then they gave vent to their feelings.

For a whole week, we lay in camp idly listening to the occasional discharge of artillery or volleys of musketry at the front. We could appreciate the rest, for the arduous duties of the past few days had been a great strain on our systems.

General Kearney was a very strict disciplinarian and never overlooked an irregularity or countenanced any crookedness. Notwithstanding, a little incident occurred in this camp that was very unlike Kearney.

Two sutlers came into camp with each a four-horse wagonload of provisions to sell. Kearney ordered them to pay the camp license before they began to sell. Nevertheless, they began to sell without paying the license, and Kearney immediately confiscated their goods, wagons, horses, and all. The goods valued at three hundred dollars. Knowing full well that soldiers would steal anything eatable they could lay hands on, Kearney sent his hostler alone with one load to the 3rd Michigan and his colored cook alone with the other to the 2nd Michigan, each having instructions to sell the goods and turn the money over to the hospital fund of our division. How the hostler disposed of his, I never knew. The cook drove his load into the 2nd Michigan camp and began to sell. As one could not wait on the customers fast enough, several of the soldiers volunteered to clerk for him. At the end of two hours, he drove the team back to Kearney and reported, "General, I done gone sol' dem goods and 'fore God, I got but two dollar an' half." Said the General, "Just as I expected." You let those rascally Michiganders steal all that stuff. Well, turn the team over to the quartermaster."

I believe none of the boys was injured by the stuff save one sergeant. He stole a whole firkin of butter from the wagon. Being unable to hide it, he was caught, court-martialed, and reduced to the ranks as a warning to others not to steal what they could not hide.

June 10 – Broke camp and moved southwesterly across the railroad and stage road, passing diagonally across the late battlefield. The ravages of shot and shell could be marked by huge limbs torn from the trees and the numberless scars on their trunks. Mounds of earth at different points indicated the burial places of the slain. A foot protruding from one and a mortified hand from another, showed that in the haste to get the dead out

of sight, the trenches had been filled so full that the recent rains had washed the dirt away sufficiently to partially expose them to view. Scenes of the last few days brought full appreciation of Bryon's lines:
"

The earth is covered thick with other clay,
Which her own clay shall cover, heap'd and pent
Rider and horse, friend, foe, in one red burial blent."

These lines had formerly seemed somewhat strained, but here they were literally correct.

Taking our place on the extreme left of the army in the foremost line, we pitched our ponchos in the edge of a wood facing an opening, and called it Camp Lincoln. From elevated points, we could see the church spires in Richmond about four miles distant. Our left rested on the edge of White Oak swamp, and the army stretched away to the right for twenty-five miles. Our brigade was to protect the flank of the army, and our regiment was to do most of the outpost duty, while the others threw up earthworks.

June 11 – Our regiment went on picket in the dense undergrowth that covers the ground along Charles City Road at the head of White Oak swamp. To gain our position, we followed a wood road nearly to the main road and then deployed right and left, forcing a passage through brush that was so thick that to see into it more than a rod was impossible.

Having established the desired line, part of the men were posted in advance to keep close watch, while the others, with knives and hatchets, cleared a space a rod wide along our entire front. The pickets were then brought back across this space under cover of the undergrowth. The enemy could not steal upon us without exposing themselves in crossing this open space. How far out there in the brush the enemy were, we knew not, but that we must not let them catch us napping we very well knew, for if they surprised us, the flank of our army would be turned and disaster follow. We were intensely alert and wore our trappings all the time, night and day, and rifles were always in reach.

The better to keep an eye on the enemy and know of his movements, Kearney ordered that sergeants with squads of men be detailed to penetrate the woods far enough to discover the enemy's pickets. These were to be sent out three times a day, morning, noon, and night. It became my duty to take the morning watch of this service; every morning we were on picket in front of Richmond.

As we worked our way through the bushes on our first trip, we came upon a log cabin in the midst of an acre of cleared ground. Concealed about this cabin all the men but three were left as a reserve. Cautiously pushing our way perhaps a quarter of a mile farther, we came in sight of the rebel pickets not far from us. Keeping our bodies as close to the ground and screened by the bushes, we made our observations without discovery.

Disliking to leave them so informally, we concluded to give them one little volley as a sort of good morning salute. Instantly following the report of our guns, there was a hasty rattling of arms and then a desultory discharge of musketry, followed by the long roll upon many drums in the distance. Satisfied that we had aroused the whole rebel camp, we crept back to our reserve at the cabin, and with them reported at once at our picket line. A similar scene to this was to be enacted three times a day while we lay in front of Richmond.

June 13 – Being relieved from picket to draw rations, we returned to camp to find that earthworks had been thrown up along our front.

Heavy cannonading was heard at the right, but it died away early in the day and although our proximity to the enemy made us alert and anxious, we soon settled down to get all the rest we could.

June 16 – Camp was alarmed at three o'clock in the morning by volleys of musketry at the outpost. Although the whole camp was profoundly slumbering at the time of the alarm, in five minutes every man was in his place in line of battle, armed and equipped for action. Before daylight, all was quiet, and we were ordered to relieve the picket line.

June 19 – Two of our men came in from observing the enemy's lines with slight wounds.

June 20 – We had been notified that the enemy was thought to be moving to the left. After a watchful night, I went on my regular turn to look for the enemy. Leaving most of the men as usual at the cabin, we moved very cautiously, intent upon discovering if the enemy was moving any troops and slightly separating from one another, we were right on their line unawares. The three men that were with me being but little separated, found themselves as close quarters with a rebel picket post. Clubbing their muskets, they struck out as well as they could and jumped back into the bushes just in time to escape a volley of balls that cut the twigs all about them. Then delivering their fire in the faces of the rebels, they saw some of them fall. Without stopping to see results, we fled to the rendezvous. Examination showed one of my men had a slight flesh wound in the thigh.

June 21 – We were relieved after having five consecutive days of this hazardous bushwhacking duty.

June 22 – Camp was in a state of alarm all day from musketry at the outpost. At two o'clock, we were marched to the rifle pits and remained there until night brought quiet.

June 23 – At daylight, firing in the rear roused the camp. We were becoming so accustomed to alarms at all hours of the day or night that the soldiers sprang up and armed without orders. Soon we were informed that Casey's division had just fired off their muskets preparatory to cleaning them. Quiet reigned in the camp and each individual soldier gained all the rest he could in expectation of his future needs.

CHAPTER 7 – JUNE 25, 1862

Once more, we went down into the dense brush to do picket duty in front of Richmond. We had hardly relieved the picket and got settled before a brisk engagement was going on just to our right. Hooker, being desirous of strengthening his picket line with the aid of Kearney's division, pushed the enemy and met with stubborn resistance. A spirited engagement took place that lasted two or three hours, in which about fifty men were killed and about four hundred wounded on each side. This battle was so near us that we could see part of the fighting from the edge of the wood in which we were. This was called the battle of Oak Grove or the Orchards, from the fact that most of the fighting was done in an oak grove and in some orchards.

Hooker and Kearney were successful, as usual, in gaining their point, and the enemy was forced to yield the ground. Musketry firing was heard at intervals during the night and occasionally the enemy sent up signal rockets. Although we followed the strict military rule of relieving sentinels every two hours, the enemy was so restless that none of us could sleep, and we passed a night of watchfulness.

June 26 – With the coming of day, while we were enjoying the bright sunshine and the pure morning air, a little incident occurred that illustrates Kearney's habit of going as far to the front as his men can go. I was standing by a tree beside the wood road when General Kearney, with a small bodyguard, rode up. Returning my salute, the general asked, "Sergeant, have you a man who has been out on this road to the enemy's lines?" Upon my answering in the affirmative, he desired me to let the man show him the enemy. Directing his escort to dismount and remain where they were, Kearney, with the soldier walking beside his horse, disappeared up the road toward the enemy. As they went, the general plied the soldier with questions concerning the enemy's position and also what he had seen out there in the bushes. As they neared an abrupt bend in the road, the soldier told the general that it was unsafe to go around the bend.

"Have you been around there?" **[Kearney]**

"Yes."

"Well, then, I want to go."

The head of Kearney's horse had scarcely shown around the bend before the rebel bullets came so thick as to cut the leaves off the trees in a perfect shower. The general laughed and said, "I guess we better back out." Wheeling his horse, they got safely away. Returning, he thanked me and said, "That's a good man you gave me. He showed me all I cared to see. I guess what you fellows haven't seen, isn't worth seeing."

About four o'clock in the afternoon, we heard terrific cannonading far away to the right. The sound indicated an important engagement, and it continued until dark. In the evening, we heard far away, faintly but distinctly, three cheers. Then it seemed to be a little nearer. Then we heard the cheers more distinctly and strains of music seemed to be blended with them.

Nearer came the cheers and music, and still nearer, until it seemed like a mighty wave borne upon the evening air from camp to camp, until it finally ended in deafening cheers and pealing strains in our own camp. The striking cadence of the Yankee cheer and the grand musical swell that swept down that long line that evening, I never shall forget. From the general rejoicing, we concluded they must have had news in camp. Planted as we were, in a swamp at the extreme end of the army, the news we received was very meager and unsatisfactory. (Days after, we learned the magnitude and result of the battle of Mechanicsville, the sound of whose artillery we heard as we guarded the left.)

[The battle at Mechanicsville, Virginia, marked the first major engagement of the Seven Days Battles.]

June 27 – With the coming of daylight, the din of battle was heard away to the right. (Several days later, we learned that this was the battle of Gaines Mills.) **[The Confederate army won the Battle of Gaines Mills, which ended with heavy casualties.]**

At eight o'clock, we were relieved from picket duty and returned to camp. At twelve o'clock, the pickets who relieved us were driven in, and we were ordered out to man the entrenchments. During the rest of the day, and until ten o'clock at night, we remained in the works listening to the distant battle, while the enemy in our immediate front kept up a scattering fusillade.

June 28 – Camp was alarmed by heavy firing much nearer than that of the two previous days. Spontaneously, all became activity, and almost instantly, we were in the works ready for action. Again, there was no fighting on our part of the line, and in the after part of the day, the sounds of strife died away.

Orders were issued to be ready to march at a moment's notice, each man to carry 100 rounds of cartridges. Speculation was rife as to our probable destination, but reliable information was kept from the troops. That we were expected to be vigorous service was evident from the amount of ammunition we were required to carry. Ready and willing to do our part, and intensely anxious, we passed the day in quiet expectation. When night drew her sable curtains down, each man retired to rest with his trusty musket for a bedfellow, expecting to need it at any moment, and each had his equipments so placed that he could buckle them on instantly.

June 29 – Pursuant to orders from brigade headquarters, we left Camp Lincoln, not to return. Early in the morning, we were on the march, taking a road that led to the rear of our camp. Passing through the timber, we came to an open field where the division of General **[Darius Nash]** Couch had been encamped. Here we halted, and two companies were detached to occupy a little earthwork at the edge of the woods, and three companies were sent farther back to a sawmill to keep a sharp lookout for signs of the enemy. The other five companies were to act as a reserve to both these detachments. While waiting in this position, we could see a continuous line of our troops defiling down the road to the rear, all marching away from Richmond. Our position and the movements of the troops, slowly forced the idea upon us that our army was retreating, and that we had been defeated in the battles of the last three days. Could it be possible that we must give up the ground that cost so much to gain? We had been told nothing, yet the evidence showed that it was even so. The few brief words spoken indicated that we had all reached the same conclusion, and every heart was sore. It was wonderful to note the change that was stealing over this usually gay and light-hearted regiment. Spinal columns could be plainly seen to stiffen, muscles became tense, shoulders braced back as if they were in stays. Jaws were firm set and ridged, and lips were drawn so tight that they became thin and white. There was a deadly gleam in those bright eyes that showed the universal motto to be "Forward, not backward." Though disappointed, they were not discouraged, and should the opportunity ever come, they will help to win back what has been lost by extra vigilance, untiring effort, and extraordinary deeds in battle.

At two o'clock, the companies on duty were called in and the regiment joined the brigade, which in turn joined the grand march to the rear on the road to Jordan's Ford. Kearney's and Hooker's divisions were marching side by side in a narrow road through the woods, intending to separate when they reached the forks of the road. Before the divisions separated, some artillery that was being hurried to the front broke the column, and cut off the four rear companies of our regiment. Not knowing what had happened, the remainder of the regiment marched on and left us. As the whole army was on the move, and no one seemed to know our destination, we were lost and could not unite with our regiment. Major Dilliman took command of these companies and made the best of his way to Jordan's Ford where he expected to find the regiment. Not finding them there, he drew us back out of the road to await developments. We had not been here long before General Kearney rode up and ordered the major, "Take your command across Jordan's Ford to Charles City Crossroad, and hold it at all hazard." We were soon waist-deep in the water, making our way across the ford. Reaching the opposite bank, we formed in line of battle and advanced rapidly through the timber to an opening. As we approached the edge of the wood, the enemy opened a heavy fire upon us that checked our advance. We were about to charge across the open field when Kearney appeared and ordered us to stop where we were. A brisk fire was maintained on both sides for some minutes during which Kearney and Birney were actively looking over the ground. Presently, Kearney came and ordered, "You will remain here as the rear guard of All Creation."

Then the troops, who had followed us across and formed in our rear, began to recross. When all were safely over, Kearney came in person and withdrew us. We recrossed, pushed by a piece of light artillery the enemy had got to the front just in time to give us a parting salute. As night was close at hand, we took the road leading down White Oak Creek, at the rear of the army, still the rear guard of "All Creation."

We moved on in total darkness, except when we passed a pile of burning baggage, and late we reached Bracket's Ford, three miles below Jordan's. The ford had been transformed into a causeway by placing big logs crosswise of the road. It was intended to chink these up with small timber and cover all with dirt, but this had not yet been done. Crossing this in the dense darkness, the four men of each file had to lock their arms firmly together to keep one another from falling. It seemed a long way over, and we got many a bump and bruise, but finally we reached the even ground and bivouacked in the thick brush a mile or more from the ford. Tired, lame, and hungry and depressed by reverses, we pushed our way into the tangled ticket and lay down to rest. Nothing but devotion to duty will keep a tired soldier from sleep. Fatigue, hunger, pain, and regrets were all soon forgotten in slumber. We had enjoyed our needed rest I know not how long, when we were rudely awakened by a terrible clatter, cracking of timber and an indescribable tearing and rushing sound, as if all the elements of destruction had been let loose upon us. In the black darkness, no one could see what caused the commotion, and all the men sprang to their feet, adding the clash of arms and their shouts of inquiry to the terrific din. The bushes were so close together we could not run, so we were forced to stand and wait the issue. After a time, a powerful voice sang out, "It's only a span of mules in a bumble bee's nest." Reassured, we lay down and slept the night out without further disturbance.

June 30 – With the first light of day, we were astir, and after a hasty breakfast, the regiment was united and joined the brigade. The brigade moved some miles to the front on Charles City Crossroad, and halted on the edge of an open plain. Details were made for picket duty, and it was currently reported that we would camp here. Some of us were so confident that this was true that we went down to a little creek among the trees and took a refreshing bath with great deliberation. The sun was shining brightly, and everything was peaceful and still, until about one o'clock when firing was heard at the outpost on the opposite side of the plain. Immediately, after the first shots were heard, we could see by the two lines of white puffs of smoke that our pickets were falling back and that the enemy was following close after. Quietly and without hurry, the brigade was got into ranks and when the pickets were all in, the command came, "By the left flank, double quick march."

Taking a road that led diagonally to the left and rear, the double quick was not slackened until we came upon our enemy in line of battle behind a low breastwork of logs and rails that had been hastily thrown together. A space had been left in the line for us, and we had hardly got into it when a terrific fire of musketry and such artillery as could be got in position was opened from both sides all along the line. Thirty thousand muskets on a side kept up their deafening music, while the artillery put in their thundering bass accompaniment. About three o'clock the 2nd Michigan was detached from the brigade and moved some distance to the right to support the 20th Indiana, who were holding a slightly constructed rifle pit, under a galling fire in the edge of the woods near a large opening that stretched away to the left and front. We closed up to the 20th and lay down behind the low breastwork in the thick undergrowth of whortleberry bushes. In a few minutes, we were buried out of sight by leaves and twigs cut from the bushes by the bails of the enemy. Through this storm of leaden hail, General Kearney, accompanied by a single orderly, rode between the lines in full view of both parties. At sight of him, our men ceased firing and began to cheer. Dropping the reins, he raised his hat with one hand with as much politeness as if he was on a review in camp. Like performances were peculiar to Kearney. A little later with a sergeant of the 20th Indiana, I went to the edge of the opening to take a view of the field.

Just at that moment, a rebel regiment emerged from the woods on the other side and charged toward a battery located on a little knoll. The battery was working for dear life, but the grape shot they were pouring into the foe did not check their advance. The fearful gaps made in their ranks were promptly closed up, and on they pressed, unwavering and in perfect line. The rebels were within a few feet of the guns when a Union regiment, lying on the ground behind the battery, rose like one man and charged the advancing enemy. Above the deafening roar of battle, we distinctly heard the clash of arms when they met. Without wavering, both came up firmly to the meeting. Just one clash, without a pause, the Confederates were borne rapidly back nearly to the woods. There the Union regiment faced about, and at double quick, retired behind the battery and laid down. The movements of both regiments were beautifully executed, and our admiration was about equally divided between the two. We felt that we had witnessed an act that is to be seen but once in a lifetime. The sulphurous smoke of the guns had shut out the light of the sun, and as the day wore away, the ominous sounds on each side of us indicated that both the right and the left had been driven back. The fighting was nearly all around us, but we could see that Kearney's and Hooker's divisions were still holding their own, and faith in these commanders and their troops kept our courage up. Night at last closed in and stopped the bloody conflict.

After the firing ceased, we were detailed for picket duty. We moved to the front and deployed across the battlefield among the dead and wounded. Language is inadequate to describe the horrors of this dark night. Men with torches searching for the wounded flitted about casting distorted and ghostly shadows upon the black pall that overhung the field. Groans, shrieks, and mad curses from the mangled wretches, yet uncared for, tortured the night air and made the surrounding gloom a fair type of Satan's Kingdom. Piteous cries for friends to come and get them arose from sufferers in every direction. One poor boy, in delirious frenzy, calls for his mother; another calls for his wife. Others call for dear friends, but most call for the comrade of their own regiment. Twenty-four different rebel regiments were represented in our front by the calls for assistance. Here amid all this hideous misery, we recognized the fact that a soldier's first duty is vigilance. A force was seen moving to our right. A strict watch disclosed twelve stands of colors, indicating a like number of regiments.

July 1 – At two o'clock in the morning, the order was whispered to us to fall back. Without a sound, we crept away to find that we had been left alone in the face of the enemy and that in the battle of yesterday—called by the various names of Glendale, Frazer Farm, White Oak Swamp, and Nelson's Farm—our army had been bent back like an oxbow, and that we must get out of the open end of the bow before daylight.

[The Seven Days Battles, fought June 25-July 1, resulted in heavy losses for both sides. Total causalities are estimated at 15,500. The Union army is forced to retreat back to Washington after Lee's Confederate forces attack McClellan's Union army.]

Noiseless as possible, we took our way to the rear. Joining some other troops, we marched side by side with them along the road through the dark woods. Presently, there was a commotion in our rear, then a rush, and I was lifted from the ground and pitched head first into the bushes. A horse and ambulance dashed madly past and then silence again reigned. I had dropped my gun in falling. Groping about, I found it in a minute and stepped into the road. Strange troops were passing. Inquiry for my regiment—no one knew where it was. In one minute, I had become lost. Following the tide, I found one of my company also lost, where a wood road branched off from the main road. While consulting with him as to our future course, the well-known forms of Philip Kearney and his horse loomed up before us and turned into the by-road. Hailing him, I asked where we could find the 2nd Michigan Infantry.

He replied, "Follow me and you'll find them." Then, alone, he dashed away into the night. In twenty seconds, the sound of his horse's hoofs died away in the distance, and we were dubious about following him on foot, but decided to take that road. With no delays, we measured off the ground as fast as we could, and when daylight came, we discovered that our road led us among the dead on part of the battlefield of yesterday. As all the dead were Confederates, we had grave doubts, but believing that Kearney would not mislead us, we kept on and soon came out of the woods in sight of some hills, covered with meadows and fields of wheat in shock. When the sun came up bright and glorious, we were upon the first ridge of the hills.

Here we halted to make a cup of coffee and broil some bacon for our breakfast. These beautiful cultivated hills were enchanting to us after having been so long in woods and swamps. As we drank our coffee and ate our hardtack and bacon, we gazed back over the road we came, admiring the rolling ground, alternated with green and gold and bordered by the woods we had left. Presently, troops were seen coming out of the woods. As we looked, their numbers increased, until they showed in dense masses in the open field.

Was it our own troops or was it the enemy? We had not seen a living soul since we turned into the by-road in the night. Possibly, we had got ahead of our army. At any rate, it was discretion to move on. As we came up over the crest of the hills, we saw troops and Kearney's battle flag a mile or more away near Haxall's Landing on the James River. Making straight for the flag, we met our regiment with other troops advancing when about half way to it.

Taking our places in the ranks, we went to the front and in half an hour, we were supporting the artillery that was hurling shells into the enemy's ranks, who occupied the ground where my comrade and I had eaten our breakfast so complacently.

Berry's brigade was assigned to support the batteries on the crest of Malvern Hills. Six full batteries, thirty-six guns, were arranged in the form of a crescent commanding a wide ravine that sloped up to the crest, while at the same time they could be brought to bear on the lower hills and woods beyond. From nine o'clock until twelve, we lay on the ground under a scorching sun behind these batteries while they kept up a constant fire, while nine siege guns, directly in our rear, rained shot and shell upon us like hail, plowing through our ranks, tearing up the ground and knocking the wheat shocks to pieces. Three mortal hours of inaction, with no chance to return the fire, under a scathing fire, calls for the utmost human endurance. The battle was mainly an artillery duel, although the infantry did hard fighting in other parts of the field. On our part of the line, the rebel General [John] Magruder, with his division charged up the wide ravine leading to the top of the hill, hoping to capture our batteries, but our artillery mowed such fearful gaps through his massed men that he failed to get half way up, and he was repulsed with terrible slaughter without any aid from us. About twelve o'clock, we were drawn a little back under the brow of the hill, where we would be partially protected from the shot of the enemy. Here we had a fine chance to note some of the antics of the shots that were plunging all around us. After a cannon ball has met with resistance in its flight, it can be seen. A shot would strike the brow of the hill in front of us and bound over our heads, lighting a quarter of a mile in our rear. Some of them would make two or three short skips while others that were rolling along about ready to stop would keep on for a half mile. An incident occurred near us that illustrates the cool stoicism of an old soldier.

Standing on the wheel of a wagon, loaded with fixed ammunition, was an old ordnance sergeant with three enlistment stripes on the sleeve of his jacket. He was directing a number of men which boxes to take to different batteries. As he bent over the wheel to point out which box he wanted taken, a cannon ball carried away the clasp knife in his belt at his back. Without looking around, changing his voice, or losing sight of the business in hand, he threw his hand back and exclaimed, "There! I've lost my knife," and the distribution of ammunition went right on.

A rebel battery was observed to have taken position in a piece of timber to our right and front, so far away that the naked eye could barely discern a movement at the place. Thompson's battery was ordered to dislodge them. Ordering his gunner to load with solid shot and aim at the spot, with telescope to his eye and watch in hand, he measured the distance by the flight of the shot. Then directing how many inches long to cut the fuse, he ordered a shell thrown at the same place. Looking for a moment through his glass, Thompson said, "That will do. You have knocked off one of their wheels, and they are getting out of there."

Just before night, several batteries arrived by the river, and they were put in to relieve the tired men who had worked their guns all day. These fresh gunners opened up a lively cannonade, which was continued long after dark, in all directions in the air, and the enemy was pushed back three miles. This ended the battle of Malvern Hills, the last of the seven days retreat. (Losses in the seven days retreat: Union – Killed 1,582; Wounded 7,209;

Prisoners 5,958. Confederate – Killed 2,820; Wounded 14,011; Prisoners 752.)

July 2 – At four o'clock, we were on the march. The men had become extremely worn out by the exacting duties of the past week. Fighting by day and marching by night had exhausted them so that they moved along mechanically following where they were led. They were so much in need of sleep that at the least halt they would sit down and fall asleep in a second. Some even slept as they marched along, being kept in place by their comrades. At one of these halts, I lay down on the ground and fell so fast asleep that I was left behind and when I awoke, the day was far advanced. I was without food and hungry and stupid from excessive fatigue, but dragging myself forward, I followed the direction of the marching troops. About ten o'clock, the rain began to fall and the roads soon became beds of sticky mortar. Thoroughly soaked, fagged out, and demoralized, I found the regiment about four o'clock p.m. with the army on a broad level plain near Harrison's Landing on the James River. Wet by the rain, the incessant tramping of the soldiers had mixed the loamy soil into thin slush, as deep as the soil had been cultivated. In this mortar bed was our camp, without one inch of dry ground or any material to keep us out of the mud if we desired to lie down. Strengthening myself with some coffee and hard bread, I went in search of material for a bed. After traveling far and searching long and faithfully, I at last got possession of a single rail. Carrying it to camp, I laid it on the ground, wrapped my blanket around me, and laid my spine along the rail. In sleep, I was soon oblivious of hunger, fatigue, pain, adversity, war, and all other ills that harass mortal man.

July 3 – I awoke to find my rail had settled so that my body was more than half buried in the mud. Despite adverse circumstance, I arose refreshed. Thanks to a vigorous constitution and the buoyancy of youth, I was ready for whatever duty might be allotted to me.

At one side of the plain on which the army was encamped, ran a wide and deep creek, making an abrupt curve around a high point on its opposite bank. On this point, overlooking our camp, the rebels posted a battery of field guns, and opened upon us at short range. The whole army fell into ranks and stood ready, while murmuring was heard on every hand that no move was being made to chastise this rebel audacity.

For some minutes, they continued to shell us vigorously, but their aim was so high that the shots all passed over our heads. Presently, one of our regiments that had made a slight detour appeared in the rear of the battery and captured it. A very cheap victory as no one had been hurt and the killing of a mule was all the damage. In the afternoon, we moved camp about two miles back from the river onto high ground.

July 4 – We quietly rested in camp. We had burnt powder enough and made noise enough the past week to satisfy the average American youth without any special effort being required for the celebration of Independence Day.

July 8 – President Lincoln inspected the army by simply riding informally through the camps. He was greeted everywhere with enthusiastic cheers.

For six weeks after the battle of Malvern Hills, we lay in camp going through the regular camp routine, drilling and reestablishing discipline in our demoralized ranks. One review being the only relieving incident. There were other troops outside of us that did the outpost duty, and we were confined strictly to the monotony of camp life. The forenoons were devoted to company drill and the afternoons to brigade and division movements.

General Kearney was always riding about upon his horse, overlooking his division during drill hours. After a few days, the monotony became irksome and discontent began to creep into the idle brains of the soldiers. The feeling was growing that we were not being properly handled, that we were losing ground that had cost much blood and treasure to gain. The newspapers that reached us were full of accounts of the interference of politicians with military movements and of the petty jealousies among our generals. As anything was better than inaction, it was learned with pleasure that we were about to move.

August 15 – Breaking camp, we moved on the Charles City Road, passing some fine plantations that had not been ruined by the army. The sight of fresh fields was a relief to us for anything in the vicinity of a large army soon becomes desolate. About noon, we passed Charles City County. From the high-sounding name, this might be expected to be a large place, but the only buildings in sight were the inevitable slave pen, the keeper's house, and a blacksmith shop. These were clean from recent whitewashing on the outside, but the inside we did not examine.

Pursuing our way, we camped for the night on the Chickahominy, where a long bridge spans it near its confluence with the James River.

August 16 – We remained in camp while the baggage, artillery, and most of the army were pushed ahead.

August 17 – We were told that we were the rearguard of the army, and that our safety depended on our reaching Barhamsville, thirty miles distant, before night. We were notified that stragglers were sure to be taken prisoners. When the sun was well up, we crossed the bridge and marched away at the swinging gait of the route step. With only two or three short halts to nibble some hardtack, we pushed on that hot August day and camped on the common at Barhamsville with the sun an hour high.

Not far from camp was a field of corn with its beautiful green leaves and silvery tassels, the first we had seen in a year. Of course, it was guarded, but I promised my messmate green corn for supper. Approaching one of the guards, I asked him to let me get half a dozen ears. As I expected, no amount of coaxing or palaver could corrupt him, but taking an opportunity when his attention was turned away from me, I leaped the fence and dashed into the corn. Securing half a dozen ears about my person, I opened negotiations to get out. The sentinel assured me he would shoot me at sight if I did not give myself up. After parleying some time, I boldly approached the fence and dared him to shoot. With angry haste, he cocked his musket and leveled it across the fence. I fell on my face, and the charge cut the corn leaves above me. With the crack of the gun, I was over the fence and away down a ravine that hid me from his sight before he could reload. We had our corn for supper.

August 18 – We camped in sight of Williamsburg.

August 19 – We marched through Williamsburg, and the deep mud that we found there in May had dried up and become pulverized into dust into which we sank ankle deep, and every step threw up such a cloud of dust that we could not see the houses from the middle of the street. We reached Yorktown and camped before night on a high plain overlooking the York River. As soon as they were dismissed from the ranks, the soldiers made a rush for the river, stripped off their clothes and plunged in. It was a grand sight to see ten thousand heads floating on the clear water of the river as they took a refreshing bath after a dusty march.

August 20 – We embarked on steamboats, and as fast as a steamer was loaded, it dropped off into the stream and came to anchor.

This ended the peninsular campaign of 1862, and this withdrawal of troops restored to the enemy all the ground we had taken from them.

CHAPTER 8 - AUGUST 21, 1862

As we lay at anchor about two o'clock in the morning, with a strong incoming tide silting up the river and the soldiers profoundly slumbering, wrapped in their blankets upon the decks, the great whistle of the vessel blew to warn all loiterers that we were about to sail. At the first note of the whistle, the soldiers, mindful of the many night alarms of the past campaign, sprang to their feet. Three men of those on the hurricane deck made a clean leap into the river, and the tide rapidly bore them away. Instantly, the cry rang out in the darkness, "Man overboard." The quick ear of General Kearney, whose quarters were in the after cabin, caught the cry and rushing to the stern door, he leaned over the rail and hailed the black darkness that hung over the water. Three eager responses came back through the impenetrable gloom. In those well-known, confident, and explosive tones, he sends back, "Paddle up, boys, we'll get you," then hurls chair after chair from the cabin into the river. There was a delay in lowering a boat on account of the lines being foul. This put Kearney in a great frenzy, and he divided his time between showering withering curses upon the captain of the vessel, pitching tables, chairs and anything that would float into the river and shouting encouragement to the floaters. After much excitement, some confusion, and a little delay, a boat was lowered, and the men were picked up and got safely aboard. When this little episode was over, the anchor was raised, and we steamed down the York River.

There were many men on the hurricane deck, and as there was no railing around this deck, Kearney, fearing a repetition of the accident just mentioned, set a staff officer to order all soldiers off that deck. There was not room enough for all to lie down on the other decks so the officer ordered the hatch to the hold taken off, and the surplus men to find room below. The provisions and supplies for the cabin table at which Kearney and staff and other officers expected to eat during the voyage, were stored in the hold. With the first light of day, before any of the officers were astir, all the men came up out of the hold and mingled with the others. When the steward went to get out supplies for breakfast, he could not find enough to feed a mouse. The officers, after some grumbling, concluded to fast that day, but seemed anxious to know who were sent below that night. As it was dark, and the men were all strange to the officers who assigned them their quarters, the case was finally dropped as hopeless. I was not in the hold, but I have ever since suspected that I ate a good breakfast out of those very supplies. One of the qualifications of a good soldier is not to ask leading question when he is presented with anything to eat.

Our voyage down the York River, across the Chesapeake Bay and up the Potomac River between the beautiful undulating ground on the Virginia side and the green hills on the Maryland shore, was a grateful diversion after the hardships and camp drudgery in the low swampy country of the peninsula.

Excepting the deep regret that the last campaign was a failure, we were happy to be once more in a delightful and healthy country.

Landing at Alexandria, we disembarked and marched through the town. The people thronged the streets and eagerly asked of each regiment, "What regiment is that?" No one would answer this question. When they asked where we came from, they received the laconic reply, "Richmond." Just outside the town, we made our camp for the night.

August 22 – We moved into town and bivouacked on a vacant lot near Christ's Church. A strong guard was posted around our bivouac, and we witnessed the going to church of the good people, the first time for many months, but we were not permitted to join them. The men were held rigidly within the line of guards so as to have them at hand when needed.

August 23 – In the early morning, we moved to the railroad station and boarded a train of boxcars in which we were taken to Warrenton Junction to join Pope's army. McClellan had been superseded in command of the Army of the Potomac by General Pope.

In a somewhat bombastic order, **[General John]** Pope, on taking the command, had promised that we should hereafter look only to our line of advance, leaving the line of retreat to care for itself.

The removal of McClellan whom the soldiers loved, and the boasting of Pope, filled the soldiers with misgivings for our future success, but they were willing to fight manfully if there was the least show for victory.

[Pope was known for having a "massive ego" that caused much strife for the Union army.]

August 24 – Moved two miles south of the Junction. The country in this section from being occupied by the troops of both armies was destitute of everything like eatables and fences. Most of the inhabitants left their homes to get away from the contending armies. Those left were the "poor white trash" and Negroes.

August 25 – The regiment went on picket duty a short distance from camp.

August 27 – We were called in from picket early in the morning and with the brigade commenced a circuitous march, following no particular road and often advancing in line of battle, as if expecting to meet an enemy at any moment. In this manner, we reached the village of Greenwich and bivouacked for the night having seen no enemy.

August 28 – Moving toward Manassas, we came to the railroad, where **[General Jeb]** Stewart's cavalry on a raid in our rear had run several trains of cars together and set them on fire. Some of the cars were still burning, but most of them were ruined. The wreck was immensely large and the destruction of property amounted to many thousands of dollars. **[General Stuart and his 1,500 Confederate soldiers destroyed supply wagons and took 300 prisoners.]**

Beside his horse, smoking a cigar under a tree on a hill, gazing on the remains of the wrecked cars, I first saw General Pope. It was natural for us to wonder if this was the way the rear was to take care of itself under this new commander. With a short halt, we moved toward Centerville in pursuit of an enemy whose whereabouts were uncertain.

Crossing Bull Run from the south at Blackburn's Ford, we traversed the bottomland and came up on to the clear plain we occupied on July 18 and 21, 1861, only this time we were faced directly in the opposite direction. On the right of the road that winds through the oak barriers that border the plain, we were drawn up in line of battle facing the woods, while two brass cannons were unlimbered in the road to our left and rear.

Passing to the front, General **[David]** Birney and staff rode leisurely along the road into the woods. Presently, a single pistol shot rang out from the timber quickly followed by a lively clattering of hoofs. The next instant, the staff officers came flying from the woods, closely followed by Birney, leaning forward with rowels deep in his horse's flank, pushed by a half dozen rebel cavalry. The rebel leader's horse's nose was at the tail of Birney's horse. With drawn saber, the rebel was reaching far ahead, goading his steed to a desperate effort to bring him a little nearer to the flying Union general.

Within twenty feet of the left of our regiment swept this little cyclone, down straight to the cannon, convulsing our boys with laughter. The Confederates, seeing their situation, wheeled their horses and were away like the wind. As they passed our flank, we gave them an oblique volley, but the spasm of laughter was not yet over and not a rebel was touched. All this did not occupy the span of more than a minute. Just at this time, General Kearney rode up and ordered us to advance through the timber in line of battle, himself riding alone in advance.

When we had pushed our way through the thick brush to the open ground about Centerville, we beheld on the crest of a ridge halfway to Centerville, the setting sun illuminating his handsome figure as he sat upon his noble gray horse, waving his hat for us to come on. As we entered the deserted rebel works, a little cloud of dust to the northwest indicated the retreat of the enemy's rear guard. We were the advance of the army that now came up and bivouacked about the town.

August 29 – Morning dawned bright and beautiful. Early all was activity as the troops moved out on the various roads that cross Bull Run as it circles around the heights of Centerville. Our regiment took the lead to the northwest and struck Bull Run at a point where the bank was high and fringed with small timber on our side, while across lay a stretch of bottom with cornfields beyond.

Deploying as skirmishers, we crossed the run and advanced across the open ground towards the cornfields. Suddenly, without warning, a rebel battery, masked behind the corn, opened a heavy fire of grape and canister upon us at short range and at the same time, a battery behind a rail fence on our left opened an enfilading fire of shell and shrapnel. Finding the enemy in force, and being in a position where we would soon be annihilated, we retired across the run to the high bank under the screen of small timber. Here we were ordered to remain as flankers, we being the extreme right of our army. This position we were to maintain throughout the battle.

While we had been occupied with this move, we could hear the deafening roar of battle all along the line. All day we vigilantly watched, while the enemy endeavored to drive us out by vigorously shelling our position, and a battery in our rear on a hill, was firing over our heads, many of whose shells, from defective fuses, burst among us. Night at last stopped the battle for the time, and we rested as we were.

August 30 – With the day came a renewal of the fight. That the field was stubbornly contested, we knew from the constant discharge of artillery and the steady roll of musketry. Of how the day was going, we were ignorant. Rumors reached us that Fitz John Porter was holding his Corps aloof from the battle, and his conduct was canvassed with bitterness. As the day wore on, the sounds indicated that the center of our army was giving ground, and our anxiety became intense, as the fighting seemed farther and farther in the rear of our position. At last, when the sun was far down the west, the remaining deployed as skirmishers to protect the flank of our barrier between the rebel cavalry and the fleeing mass of constant fire upon us, except when their cavalry intervened to charge us.

Under their galling fire and the continual menace of their cavalry, we had to hold off the stragglers from breaking our line and demoralizing our men. Every fugitive had a frightful tale to tell. Ours was a heavy responsibility. With our feeble line broken, the enemy could ride in and capture prisoners by the thousands. As we were crossing a stretch of bottomland, we came to a creek that crossed our line of march. As we were fording this with half the men on one side and half on the other, a squadron of cavalry charged us in fine style, thinking to take us at disadvantage. Their leader rode in advance, shouting "Surrender!" at every leap of his horse. With the coolness and precision of an afternoon drill, our boys rallied by fours, shouting with all their might and delivered a withering fire directly in their faces. Their horses swerved to the right and left and went flying madly to the rear, many of them with empty saddles. Battle of Second Bull Run – 2nd Division loss 14,800; Confederate loss 10,700.

[The Battle of Second Bull Run, or the Second Manassas, August 29-30, 1862, once again resulted in a Union defeat after a major miscalculation by General Pope. Present day records state more than 22,000 men were killed or wounded in this battle.]

That was the last annoyance we had from the cavalry. The artillery kept up their fire, but they were beyond our reach and the effect on their line was slight.

Mounted upon a superb bay charger, a woman rode furiously among the male rout, calling up the able bodied among those wounded, bleeding, hatless, coatless fugitives from the battlefield to form ranks and repel the foe. She enforced her demands by appeals to their manhood by calling them sneaks, cowards, villains, and traitors, and when both these failed, by a look into the muzzle of her navy revolver. Who she was or what became of her, I had no opportunity to ascertain.

There was a pressure all the time, from men who had lost their own command, to take their places in our line. One of these, a captain, took his place in my section of the line and began chattering of the terrible things that had happened to his company. I promptly ordered him out of the line, when he disdainfully asked what authority a sergeant had to command a captain. Dropping the point of my bayonet to his breast, I informed him, here is my authority. My captain, seeing the trouble, came at once to the scene. Pausing only for one glance, at the straggler, he said, "Put him out of the ranks or kill him." Our hands were too full of this great emergency to be formal with officers who could not take care of their own men.

Night found us near a wood through which wound a road toward Centerville. Closing up our ranks, we pressed on and camped with the army at that place.

August 31 – We lay in camp through a cold drizzling rain.

September 1 – We left Centerville by the Fairfax Road at two o'clock p.m. When about four miles out, the rain falling in torrents, at four o'clock, we came upon a battle going on at Chantilly. The enemy had made an attempt to pierce our line and brought on a spirited engagement.

Kearney, with his customary energy, hurried his troops to the front, but the heavy rain and the darkness of night stopped the fight before we were fairly engaged.

In attempting to put a detail on picket in a cornfield, our regiment lost some men and the detail was withdrawn. Then Companies D and E were assigned to establish the picket line. One lieutenant was all the commissioned officers that Company D had at this time. Himself remaining in the pasture by the cornfield, he directed me to place the picket in the cornfield, connecting at each side with the lines outside the corn.

Dividing the company into two parts, I sent Sergeant Hardy to the left side of the field with one squad, while I went to the right side with the others. Both taking the same row of corn, we successfully placed our men and met in the center of the field. It was intensely dark and the rebel line was so close to us that a loud whisper could be heard from one line to the other. The least noise from us was sure to draw a volley from the enemy.

Our line led directly across the field, and we were continually stumbling over the dead in the darkness. With all my caution, I ran against the feet of a dead man and fell down upon him, striking my hands in his bloody face and besmearing them with his blood. Although we had drawn the enemies' fire several times, at last, our line was established without loss, and I notified the lieutenant which row of corn to follow should he wish to communicate with me. We settled down to a night of anxious watchfulness in the loose soil of the cornfield, which had been soaked to mortar by the rains of the afternoon. Our garments, already wet, were deluged by water from the corn leaves with every breath of wind. The air was chilling; the dead were all about us and to move was to invite death from the enemy's guns. A more cheerless situation could not well be imagined. Yet some slept there among some sitting upon the dead despite the fact that a single move might give them rapid transit to the other world. The inordinate strain was so great from toil and exposure that the fear of death would not keep them awake.

September 2 – At three o'clock a.m., a messenger from the lieutenant came, creeping on hands and knees, with an order to fall back to the fence at once. Whispering the order close in the ear of the man on my right and on my left, it was passed down the line in the same manner. Without audible sound, the men reached and climbed the fence into the pasture lot where we had left an army the night before. General Kearney was killed and his body and horse were sent within our lines by the enemy. Battle of Chantilly – Union losses 1,300; Confederate losses 800.

[From **Life Stories of Civil War Heroes**: Kearny's aide de camp wrote about his experience that fateful night in a piece entitled "The Death of General Philip Kearny."

"I was Aide de Camp to General Kearny, and accompanied him

during the Battle of Chantilly, when he rode on in advance of his Division to see the position occupied by the troops of General Stevens whom we were to relieve or reinforce. We rode along the line, and General Kearny sent off one staff officer after another with orders, until I was the only one left with him. We finally arrived at the right of Stevens' line, where a battery was shelling the opposite woods. The General ordered me to ride at a gallop, back to General Pope, commanding one of our Brigades, and order him to "double-quick" his brigade to that point and go into line. I did so, and returned as quickly as possible to the Battery. The rain was falling fast and darkness was coming on. I inquired of the Battery men which way General Kearny went, and they replied, pointing down to the right and front, "that way." "My God," was my exclamation, "we have no troops there, he has ridden right into the enemy lines." And so it proved. Wishing to know the nature of the ground and whether the woods were occupied or not, he rode with his usual bravery, to his death, as we learned from the Confederates, who next day brought in his body under a flag of truce. The General rode up to a whole company of the enemy, paid no attention to their demand that he surrender, wheeled his horse and started back. The whole company fired a volley, but only one bullet struck him; that entered his hip as he lay low along the horse, and came out at the shoulder. And so fell the most picturesque and gallant soldier that it was my fortune to meet during the war."

Today a statute of Kearney on horseback stands over his grave in Arlington Cemetery.]

The uncertain light of breaking day revealed no sign of troops to our astonished gaze. Instantly, the truth flashed upon us that we a mere handful of men, had been left in the face of twenty thousand rebels to hold the front, while our artillery, baggage, and troops were moved to a place of safety, and there to make our escape, if we could. Failing to make good our escape, we were to be sacrificed for the good of the army. In deathlike silence, our ranks were closed up and a retreat toward the main road was cautiously begun in lieu of battle. As soon as safety would permit, the time was changed to double quick. From my place in rear of the retreating line, the noiseless fall of vanished feet upon the soft green turf in the indistinct light seemed like fleeing shadows. In the shortest possible time, the road was reached, and the direction was changed by the flank, but the double-quick step was maintained.

The coming light had broadened to the full light of day when we slackened our pace in sight of the rear guard of our army near Fairfax Court House, four miles from the battlefield of Chantilly.

Safely within our own lines, we sought a brief rest and breakfast. Before noon, we were again on the march toward Alexandria.

September 3 – Before noon, we reached Fort Lyon overlooking Alexandria. We were back on the Potomac where we were over a year before.

Pope's Virginia campaign was over, and the enemy occupied all the ground he ever had occupied in that state.

As we approached Fort Lyon, the 20th and 24th Michigan regiments, fresh from home, with their bright and clean new uniforms, ranged themselves along the roadside to look at us. With ill-concealed disgust, they viewed our tattered clothing, begrimed with mud, ashes, pitch smoke, and grease, and disdainfully asked, "What makes you so dirty and ragged? Poor Boys." They had yet to learn the dreadful trials, hardships, discomforts, discouragements, and dangers of a soldier's life. With no chance to wash our clothes, we had been twenty days incessantly marching or fighting, pushing through tangled thickets and briars, fording streams, rolling in the mire and dust, in sunshine and in rain, until not only our clothing, but also our bodies, were filthy and covered with vermin.

September 4 – It became known that we were to take the place of troops that had been occupying the fortifications on Arlington Heights while we recruited our health and obtained new clothing. The fresh troops that had been held for the defense of Washington were pushed out to head off General [Robert E.] Lee in his invasion of Maryland.

The lieutenant, our only commissioned officer, was taken sick, leaving me in charge of the company, with the responsibility of drawing from the quartermaster almost a complete outfit of clothing, and of issuing them to the various members of the company. The lieutenant was absent ten days during which time we had all been fitted with clean clothes and had moved a short distance each alternate day along the works, and now were at Fort Ward, north of Fairfax Seminary where we were encamped from September 15 to 25. We encamped on Upton's Hill, September 25 to October 11.

The battles of Antietam and South Mountain were fought while we guarded the fortifications and recruited our strength for fresh exertions.

Our occupation was the simplest camp duties and few were the incidents to break the monotony. Fresh vegetables and some of the luxuries of life could be procured, and these were indulged in to the last dollar of greenbacks, and these failing bills on broken northern banks were used. This might seem dishonest, but there was great provocation. Farmers came inside the guards with their truck, asking fabulous prices, expecting to make a fortune out of a single load of vegetables, when they knew that the soldiers could buy at ordinary prices if they could get out of camp. One instance will illustrate the many. A man came into camp with a one-horse cartload of garden truck, which he sold out by the price of nearly four hundred dollars. As he was about to leave camp for another load, someone told him that the money he had received was worthless. Going to the colonel, he complained that the soldiers had passed bad money upon him. The colonel demanded to see the money. When it was produced, it proved to be bills on the Bank of Adrian, Michigan, and signed O.M. Poe, Cashier. The colonel gazed on the bills in astonishment and finally said, "They ought to be good for they have all got my name to them. The signature, however, is a forgery. Show me the men that passed them, and I will have them punished."

This he could not do as one soldier looks like all others to a civilian, and he was forced to depart wiser and sadder than when he came.

Our time was devoted to rest and refreshment, and we enjoyed for once reading accounts of the army of the Potomac in battle, and we not with them. Lee had been driven back across the Potomac. Fresh troops were arriving, and we made preparations to move on and leave them to hold Washington.

October 11 – With our brigade, we broke camp, taking the road to Chainbridge and crossed into Maryland. Continuing the march up the Potomac past Tinallytown, Rockville, and Danistown, we bivouacked for the night in a grand old wood near a romantic little glen, through which leaped and gurgled a rocky streamlet on its way to join the Potomac. Healthy and vigorous from our month's rest, the march over this rolling country clothed in green and amid the delightful scenery of the upper Potomac with its pure and healthy atmosphere, seemed like a pleasure trip.

October 12 – Resuming the march, Seneca Mills was reached by ten o'clock, where we formed line of battle in expectation of an attack from Stuart's cavalry. After waiting two hours, it was found that the rebel commander had changed his mind and departed for other fields. Proceeding, Edwards Ferry was reached at seven o'clock. The prominent features of this point are the ferry across the river and the locks in the Chesapeake and Ohio Canal, which runs parallel, and near to the river. Just below the locks stand a warehouse where goods are unloaded from the canal. For fifteen days, we were destined to guard the locks and do picket duty along the river. The weather was delightful; vegetables, milk, and butter in abundance could be procured from the farmhouses, and large fresh bread daily from the Washington bakeries. With these advantages, we lived upon the fat of the land and reveled in the luxury of glorious weather.

Our duties were very light, and the memories of this camp linger like those of a summer picnic.

Every camp had its incidents. Mention of one or two will suffice for this one. Beside the lock was a wash weir and in the wash weir was a contrivance of slats, called an eel pot. Those of us near the lock bargained with the lock keeper to catch eels in his eel pot on condition that he have half we caught. The old man was supplied with all the eels that he and his daughter—his entire family—could eat, but as there were several of us living upon eel, he suspected that he was not getting full half. Coming out of his house one morning and seeing a dozen men all cooking eels while he was tendered only one fine four-pounder, he demanded more. Being refused, he was very wroth and jumped down into the pot and swore he would tear it up. Wrenching off one slat, he raised up to throw it on the bank when he saw a dozen muskets all aimed at his head. In speechless astonishment, he gazed one instant, then guiltily said, "I'll put it right back." The eel fishing was disturbed no more.

One afternoon, a comrade and I went into the cabin of a canal boat that lay moored to the bank to take a quiet sleep. Presently, our nap was disturbed by a peculiar bumping of the boat. Investigation showed that in the spirit of mischief, our comrades had cast off the moorings and the boat drifted with the wind, a mile down the level. Jumping ashore, we hurried back to escape censure for being absent without leave.

October 28 – We left Edwards Ferry and moved up the river nine miles to Whites Ford. At this point, the Potomac broadens and shallows so as to be fordable. In mid-channel is a small island. Here we found troops crossing into Virginia. The current was quite rapid and the water was cold and waist deep. The troops preceding us were slow and straggled badly in getting across. Our colonel held us until nearly all those in advance were out of the water when he said, "Now boys, show those fellows how to ford a stream." Entering the water with a shout, we cleared the stream and were on the high bank ahead of some that were nearly over before we started. The banks on the Virginia side were nearly perpendicular, and the cannon had to be hauled up by ropes. After hauling up those belonging to our brigade, we moved back a short distance and built large fires of rails and dry logs to warm our limbs and dry our clothes.

October 29 – We marched toward Leesburg.

October 31 – We reached the vicinity of Leesburg, a village that nestles at the foot of a high range of hills or a low range of mountains, called the Kittoclan Range, a spur of the Blue Ridge Mountains.

Leaving Leesburg to our left, we began a gradual ascent of the hills. Taking a shortcut from one road to another, we passed directly through the grounds of Mr. Swan, the President of the Baltimore and Ohio Railroad. Passing through an arch, the uprights of which were surmounted by large cast-iron eagles, we traversed the deep park and approached the castle-like residence situated half way up the side of a mountain. The building and grounds were very imposing and the view was magnificent. Without halting, we toiled on to a still greater elevation and camped on the summit of one of the lesser hills of the range. Here we found General **[George]** Stoneman's cavalry division, the cavalry and our brigade forming a corps of observation, which was to scour these mountain ranges to discover a lurking enemy and gain information that might be of use in future campaigns. Stoneman issued an ironclad order forbidding his soldiers from taking anything from the inhabitants without pay on pain of having their heads shaved, their buttons cut off, and of being branded on the cheek with the letter "T" for thief, and drummed out of the service. Our marches were rapid, and their directions were changed from day to day.

November 1 – Standing upon Faitheys Hill, the highest peak of the range, Leesburg lies at your feet like a toy. To the north, thirty miles away, over in Maryland, the Sugar Loaf Mountain, clothed in green, looms high above the surrounding country. The Potomac, a silver thread, winds its way for miles among the forest-clad hills. Washington, sixty miles away, glitters in the sunlight. The expanse between the river and the Blue Ridge Mountains lays spread out like a panorama, with cultivated fields, beautiful dwellings, and woodlands. To the south, the crest of the range, with it s undulating swells dotted with brown fields of corn, stretches away with its diversified scenery. To the west, the Blue Ridge Mountains, towering, crazy and wild, their lofty peaks and dark glens, veiled a thin haze of blue smoke, completes a picture of grandeur and beauty.

Leaving these beauties of nature, we went on picket near the village of Hamilton on the Winchester Road.

We were now in a country that had never been devastated by an army and poultry, vegetables, and dairy products were plentiful. Thirteen dollars a month would not buy many of these things; still the soldiers indulged to the full in these luxuries despite Stoneman's withering order.

November 2 – From my post in the edge of Hamilton village, it was amusing to hear a rooster begin a lusty crow in the early morning and change it to a squawk right in the middle.

At ten o'clock, we took up our march along the western side of the Kittoclan Range. A little skirmish near Snickersville resulted in driving some rebel cavalry through Snickers Gap in the Blue Ridge.

November 3 – Our march led us along the base of Bull Run Mountains, through Mountville and New Lisbon, in the direction of Ashley's Gap. As we traversed this rough and rugged country, government rations were almost entirely discarded, and we lived off the country.

November 5 – We passed through Middleburgh and camped on White Plains. Just before dark, my tent mate located a flock of sheep. As it was not more than two miles from where we camped, we sallied forth armed with a butcher knife in search of good fat mutton. Arriving at the sheep pasture, we found several of our command ahead of us with the sheep cornered but without a knife. Finding I had a knife, they installed me head butcher. Four had been slaughtered when a mounted officer was seen riding toward us through the gathering darkness. Instantly, I was left alone among my dead lambs. Seeing but one man approaching, I stood my ground. The well-known form of Stoneman's inspector general rode up to me and briskly asked, "What are doing here?"

"Getting mutton for breakfast."

"Have you heard General Stoneman's order?"

"Yes."

"Don't you expect to be punished for disobedience?"

"Not until I'm caught."

"Haven't I caught you?"

"No, you are only one man, I am another."

Laughing, he said, "All right. Give me that lamb for General Stoneman's table and don't kill more than you need," and away he rode.

November 6 – We reached the vicinity of Warrenton.

"November 7 – Toward night the first snowstorm of the season set in and gave us six inches of wet slushy snow.

[On this date, President Lincoln replaced General McClellan with General Ambrose E. Burnside as the new commander of the Army of the Potomac.]

November 8 – The sun quickly melted the snow. We marched to Waterloo.

November 10 – We crossed Hedgermann's River on the Gordonsville turnpike and camped about two miles from the river. Our two weeks marching in a zigzag direction over the rugged hills had little of excitement, but much toil. We remained in this camp until November 15, when we were transferred from the 3rd Brigade, 5th Division, 3rd Corps to the 1st Brigade, Burns division, 9th Corps.

This was the end of our corps of observation, and our future service was to be with other troops and other commands.

CHAPTER 9 – NOVEMBER 15, 1862

Leaving our old comrades in the 3rd Corps, we joined the 1st Brigade, 1st Division (Burns), 9th Corps at White Sulphur Springs.

The springs are strong of sulphur, and at one time, many invalids sought them for their medicinal qualities. Aside from the hotels near the springs, there was no town worth mentioning.

At this point, rebel cavalry had dashed in and captured a wagon train, and our brigade remained here until all the trains were past.

The brigade consisted of the 2nd, 8th, 17th, and 20th Michigan regiments and the 79th New York (Highlanders) regiment. The 8th and 79th were old regiments and had served together in South Carolina at Hilton Head, Bufort, Port Royal, James Island, and other places. From sharing dangers together, their friendship was very strong. The 17th was a new regiment but had met the enemy in one engagement at South Mountain, while the 20th had yet to receive their "Baptism of Fire," being fresh from home. The 8th and Highlanders, having served in a small army where their valor had made them leaders, were very patronizing toward the fresh troops. The 2nd having been connected with a large army where fresh troops arrived daily, received them into fellowship as equals. These circumstances seemed to divide the brigade, in all trifling differences, into two factions. The 2nd, 17th, and 20th against the 8th and 79th. But for the general good the brigade was a unit and all joined in support of each regiment.

November 16 – Marching by a devious route, Layetteville, Liberty, and Bealton with their tumbledown houses and poverty-stricken appearance were passed, and we camped near the Orange and Alexandria Railroad on the same ground we left August 27.

November 17 – The 18th regiment taking the lead, followed by the 20th, 17th, and 2nd, with the 79th in the rear, we marched twelve miles toward Falmouth. The weather was cold and raw. Rest at night depended largely on a supply of fuel. The 8th being ahead, were first on the ground where we were to camp for the night. Stacking their arms, they quickly gathered all the rails within reach of camp and put guards on them for their own use and for the 79th, refusing to let the new regiments have a single rail. When the 2nd came up, the boys of the 20th were wondering what they should do for fire to make their supper. The 2nd took in the situation at a glance, and as soon as their arms were stacked, the cry rang out, "Rally on the rails!" Making a grand rush, some seized the guards and held them, while the others, assisted by the 17th and 20th, carried off every rail. Seeing they could not help themselves, they complained to the colonel, and he to the brigade commander. Being informed the rails were under guard, and were taken by force by that reckless 2nd Regiment, the brigadier quietly remarked that no one had authority to post such a guard but himself, and he had not ordered it. That quieted the matter, and we obligingly allowed them to cook by our fires, but they never forgave the other three regiments. The wholesale business was never again undertaken in that brigade.

November 18 – Our camp was pitched twelve miles nearer Falmouth.

November 19 – Continuing our course, we passed through Falmouth, a representative little Virginia town on the Stafford County side of the Rappahannock River, and camped on high ground opposite Fredericksburg. The enemy was in possession of the town and country on the Fredrick side and before dark, we had opened verbal skirmishes with them across the river. Among other chaff, they desired to know when we were crossing over to take the town. As all boats and bridges along the river had been burnt previous to our arrival, of course, our army could not cross without pontoon bridges.

Pontoon bridges are made of boats, anchored lengthwise of the stream, across which stringers are laid, upon which the planks are placed, forming a floating bridge. In transporting these pontoons, a boat is placed upon a wagon, and in it is placed the stringers and planks belonging to it.

Burnside expected a load of these boats to meet his army at Fredericksburg, but through a misunderstanding, they were not on the ground. Waiting for these, our army camped on the broad plain that rises abruptly, high above the river and extends back, dotted here and there with fine plantation houses. Across from us nestled the smart little village of Fredericksburg, on the bank of the river, with its railroad leading to Richmond.

The Rappahannock is quite narrow, but is navigable as far as this place. Back of the town and parallel with the river rises a succession of ridges, one above another, with slight intervals between until the crest is reached about two miles to the rear and at a high elevation above the river and Stafford side. The possession of this admirable series of defenses was the object of our moving in this direction and could have been easily taken, if we were across the river.

The pontoons were not promptly on hand, and the whole army felt that we were losing a golden opportunity. That night as we rested in camp, we saw, far away to the southwest, the faint glow of light on the sky that we believed to be the halo of an army's campfire.

November 20 – No pontoons yet. At night, the light in the southwest was plainer and nearer.

November 21 – Before the close of day, masses of troops were seen on the heights back of Fredericksburg, cannon bristled from every ridge, and at night, the enemy's campfires blazed over all that country. From that time, there seemed to be a settling down of the army to organize, and no move of importance was made for several days. Burnside divided his army into three grand divisions, commanded respectively by Sumner, Hooker, and Franklin.

November 28 – The second regiment was detailed to support Thompson's battery of regulars, posted opposite, and a trifle below Fredericksburg, near the river. In performance of this duty, our pickets were posted close to the water's edge, and the rebel pickets were also close to the river on the other side. The river being very narrow, many a friendly chat ensued between the two parties. Some sly contraband trade in coffee and tobacco was indulged in by means of small boats made of bark with paper sails. Our fellows would tie on a small parcel of coffee, trim the sails, and send it across. The "Johnnies" would take the coffee, supply its place with an equivalent of tobacco, shift the sails, and send it back.

The supplies for our vast army were hauled by teams through a swampy country from Aquia Creek and Belle Plain Landing on the Potomac. One morning, we awoke to find the ground covered with six inches of snow, the weather cold, and no breakfast. The snow had prevented teams from getting to us with provisions. There was some suffering, but this state of things did not last many hours. Personally, I overcame my hunger by paying a quarter for a pint of corn meal, mixing it with water, and baking it in the ashes.

After ten days service with the battery, we returned to camp and began the construction of winter quarters. Our army lay encamped for miles along the river and back into the country. All were actively engaged in the preparation for winter. Each company was required to keep on its own ground, and its huts must be in line, but individuals chose their own style of building. There was much similarity yet there were no two alike. The house built by two comrades and myself was a type of them all.

Cutting oak poles, we laid up the walls of a hut, plastering the cracks with clay mortar and roofing it with our poncho tents. A low door was left but no window; the dim light that filtered through the tent cloth was thought sufficient. A chimney of sticks plastered with clay completed the house and some rude bunks and a primitive writing desk was all the furniture. This was to be our winter home. The writing desk was made from a box picked up in the woods by my partner and had evidently held hospital supplies in the form of liquors. The possession of these boards afterwards caused me to attend a general court martial many days as a witness.

While the army had been getting in shape and making itself comfortable, the delayed pontoons trains had arrived. Our bridge was laid some distance below the town, and it was thought desirable to have two more near to or opposite the town. Every attempt to lay these latter bridges drew a murderous fire from rebel sharpshooters concealed in cellars and behind stone walls along the riverfront not one hundred and fifty yards from the water. The Engineers Corps, under cover of the darkness, unloaded a number of boats with their accessories in front of the town and started a bridge, but could not proceed on account of the destruction fire of this concealed enemy.

December 11 – Every other expedient seemed to have failed and at ten o'clock, Burnside gave the order, "Concentrate the fire of your guns on the city and batter it down." In a few minutes, one hundred and seventy-nine cannon of various caliber, stationed on available eminences on the Stafford side, belched shot, shell, and grape upon the doomed city. Amid the deafening roar of all these guns, we watched the deadly projectiles plunge through the buildings, rip off their sides, and tear their roofs to splinters. The iron messengers skipped broadcast through every part, carrying away portions of brick walls and church steeples, reducing the place to an unsightly ruin. This devastating fire was kept up for an hour. Each gun had fired fifty shots, and the town was on fire in several places.

During the bombardment, attempts had been made to lay the bridges, but the gunners found it impossible to depress their guns enough to reach the opposite river front, and the rebel sharpshooters remained in comparative safety. The experiment was a costly failure, and a forlorn hope was the next thing to try.

Burnside called for volunteers to go over in boats and drive the rebels out. Colonel Hall, with the 7th Michigan Infantry backed by the 19th Massachusetts, promptly came forward for this mission. Rushing down the steep riverbank, they found temporary shelter behind the pontoon boats and planking of which the bridge was to be made. When all was ready, our artillery opened a vigorous fire as nearly as possible on the lurking foe. Then the 7th rose, pushed the boats into the river, leaped in twenty to thirty to the boat and pulled lustily for the other shore. Then the rebels got in their murderous work with terrible effect, but as the boats passed the middle of the river, they came under the bank out of immediate danger. When a number of boats had come to land, they charged up the bank, and the rebels tumbled out of their snug retreats and scampered up the streets of the town, dreading to close with men that would cross a stream in the face of their fire. Those that could not make a safe retreat came out and surrendered. By this dash of genuine bravery, the problem was solved, and the bridges laid without further delay. The whole army had been ready to move any minute all day, but night coming on only enough to hold the ground already gained were pushed across.

December 12 – While a mist hung over the scene, troops were pressing over all the bridges. Our brigade crossed the center bridge and remained on the dock in the eastern outskirt of the town all day. The heavy guns on the Stafford side were playing directly over our heads and many of their defective shells burst above us, and the fragments rained down among us. The enemy kept up a constant cannonade upon us, but most of their shots curved downward into the river two hundred yards in our rear.

Franklin's division, two miles below, was fighting some, and the sounds of strife were hot above the town.

The 20th was under fire for the first time, and we were inwardly amused to see a half frightened expression come over their countenances at the terrific explosions. Then glancing at the composed attitude of the older troops, they seemed to feel reassured and accepted the situation as a matter of course. This was the introductory to a splendid career for the 20th.

Wide-awake soldiers always laugh at the ridiculous, no matter what the danger. A Negro boy of perhaps sixteen years kept the boys convulsed with laughter for some time by his antics. He was keeping close under a low bank, scared entirely out of his wits. When a shell would burst above us, he would dive headfirst into the soft mud and writhe and squirm like a mangled worm. This he repeated at every recurring shot until the officers drove him away. The sight was pitiful but very ridiculous.

In our division was a regiment of Germans, commanded by a short thick German with a ponderous belly who gave most of his commands in German. These were deployed as skirmishers and moved up the slope a few yards in advance of us and laid down. The brigadier ordered him to move his men farther out. Getting upon his feet with much trouble, he drew his sword and commanded, "Stanoup." Not a man stirred from his horizontal position upon the grass. Throwing back his shoulders, advancing his immense belly and taking a deep breath, he roared, "Up Stander, Gott Fer Dam."

It is needless to add that the men moved, and equally needless to say, that our whole brigade split with laughter and repeated the Dutch colonel's command a hundred times during the day.

December 13 – The brigade was occupied in keeping open communication between Sumner's and Franklin's grade divisions. We were in the center of the line, but the fighting was confined mostly to the flanks, so that we were not actually engaged in fighting although the enemy shelled us vigorously from time to time. Far way to the right and to the left, we could hear the sounds and see the smoke of battle.

December 14 – Our brigade was moved on to the first line of hills directly back of Fredericksburg and between it and the enemy's center. There was skirmishing along the line but no energetic action.

December 15 – Burnside ordered the 9th Corps to attempt to carry the enemy's center by assault in massed column at ten o'clock. The corps was formed in column by regiments, the 2nd Michigan in front. As we stood there and gazed up at the successive terraces frowning with heavy cannon to the crest two miles away, each man said in his heart, *none of us will live to get there*. All knew the danger, yet there was no trembling, no shrinking. With this impossible task before us, we stood ready, only waiting for the order.

At the last minute, the order was countermanded. The other generals were unanimously of the opinion that it was a useless sacrifice, and we were saved from annihilation. Toward night, a hundred volunteers were called for to go on a special service. I volunteered to go and was installed orderly sergeant of the company. General Fenton of the 8th was in command and kept us a little separate from the brigade until after dark. Then he conducted us directly to the front out through out picket line. Choosing a position behind a stone wall close to the rebel lines, he directed me to let the men sit down behind the wall and keep very quiet. Seating himself in the loose sand in the road, he requested his officers and noncommissioned officers to take seats about him. Sitting there in the dirt outside our lines and right under the rebels noses, he beguiled the tedious hours of the night by telling humorous stories in a low voice.

December 16 – At three o'clock in the morning, came creeping up to us a messenger with orders to fall back at once. Without noise or confusion, our little band was formed and turned its face toward the river. Arriving where we left our picket lines, there was nobody. Moving on to where we left our brigade, there was not a soul there. Without waiting to voice our surprise, we pushed onto the river. There we found General **[William Wallace]** Burns sitting upon his horse alone, excepting a couple of engineers in charge of the bridge. As we marched onto the bridge, the anchors were cast off, and the current graceful swung the bridge across the river to the other side. Now we knew what we "went out for to do." We went out to show a front to the enemy while the army was withdrawn across the river and then to save ourselves if we could.

As the gray of dawn crept up over the east, we returned to our winter huts to remain quiet for a season. The season of disagreeable weather was upon us, and we argued from the past that we had struck our winter home. We could enjoy the dull monotony of camp for a season with content and profit, but the long weary days to the end of the winter looked far distant.

[The Battle of Fredericksburg, December 11-15, 1862, resulted in a

Confederate victory with nearly 18,000 men killed or wounded or missing.]

December 25 – Christmas? Well, what of it for a soldier in his crowded little hut with its smoking fireplace, earthen floor, and cloth roof? Without comforts, conveniences, or accessories, he can sit himself down to drag the time away until active service shall again demand all his energies. He had shown many a time that he cared little for himself, his health, or his life. Now that he has leisure to reflect, does he care for the regard of others? Would it interest him to know that he is remembered by friends and associates of other days? To know that he is missed in social gatherings near his old home or at his home itself? At least there are no feasts for him. He eats his regular rations of hard bread, bacon, and coffee, and tries to be content. The day passes like all others with its regular round of military duties.

By way of variety with a comrade, I went to visit friends in the 4th Michigan who were camped about four miles in our rear. The day was fine, and we enjoyed our visit very much. When returning we met a group of officers on horseback. Passing right along without paying any attention to the officers, one of them turned back to demand why we had not saluted him. With the utmost promptness, we informed him that it was not the custom with us to salute every jackass we met. In great rage, he demanded our names with regiment and company. These we truthfully gave him, but we never heard from him after. He was young and indiscreet, or he would not have turned back for such a purpose.

The last week of 1862, Burnside's army lay in camp inactive.

The winter rains had set in, and it was almost impossible to get supplies for the army over the miry roads from Aquia Creek and Belle Plain Landing. With the whole surface of the country one vast mortar bed, active operations were not thought of in the army. Yet every newspaper that reached us was full of condemnations for the idleness of the troops in the field. Any attempt to move large bodies of men was inexpedient and to move artillery and supply trains was next to impossible.

Between the clamor of northern papers, the quarrels among general officers, and the interference of Congress with artillery movements, the rank and file of the army of the Potomac was becoming discouraged and demoralized. The men were beginning to feel that they were enduring hardships and that lives were being sacrificed without adequate results, because of petty jealousies among the leaders. Idleness and discontent go hand in hand with soldiers, and the gloomy outlook of our winter camp was not cheering. The fences had all disappeared for fuel and green wood for cooking and heating purposes had to be hauled long distances with the mules floundering knee deep in the mire and the wagons cutting almost to the hubs.

The few inhabitants that stayed in their houses were necessitated to ask food from the government. Amid this wretchedness with rain falling daily, nothing was done but to guard the roads to our base of supplies and keep a strong picket guard along the Rappahannock River.

An incident occurred about this time that caused considerable talk.

Directly in front of our camp and also directly opposite Fredericksburg stood an old-fashioned brick house occupied by an old lady and daughter. A young lieutenant, whose duties called him to the house, became acquainted with the young lady, and at her invitation called frequently upon her. After a few visits, a mutual attachment seemed to spring up between the young people. The lieutenant proposed marriage and was accepted. One evening while calling upon his affianced, during a brief lull in the conversation, the heavy atmosphere bore to his ear what he judged to be the click of a telegraphic instrument. Instantly, his interest and loyalty were awakened and a suspicion of treachery aroused. Without betraying a suspicion that he had heard the sound, he chatted on, his keen ear strained to catch and locate the clicking.

At the usual hour he left, convinced that a contraband communication was going on with the enemy. The next evening, taking with him a strong guard and leaving them in the yard, he again called upon the young lady. Receiving him with the warmth of an expected bride, she conducted him to a sofa and there clasped in each other's arms, they indulged in fond caresses and endearing words until the ominous sounds of the clicking telegraph again greeted his ear. Excusing himself for a moment that he might clear the phlegm from his throat, he opened the door and expectorated vigorously into the darkness. While the door was still open, the guard pressed in and exhibited an order from Burnside to search the house. Then the young lady, so recently the devoted lover, became a tigress. With flushed cheeks and blazing eyes, she let loose a torrent or rage and abuse upon the Union soldiers.

"Yankee brutes, Lincoln hirelings, scum of the North, and cutthroats" were too feeble words to express her hate. Familiar with the favorite expressions of southern ladies, the guard with due deliberation proceeded with the search. Down in the cellar, they unearthed a young man with complete telegraph offices, the wires leading underground to Fredericksburg. They brought the cringing knave up into the habitable world, and he pleaded piteously for his cowardly lie. The sight of his abject fear aroused the genuine affection of the young lady, and she begged in tears of the lieutenant to spare the life of her dear husband. A married woman, she had played lover to the lieutenant for the sake of the little information she could squeeze out of him for the use of the rebels.

There are many instances where southern women decoyed men to be taken prisoner and some to their deaths. They did not hesitate at anything if they could cripple a Yankee.

CHAPTER 10 – JANUARY 1, 1863

The disagreeable inclement weather of a southern winter was upon us. Wet, slushy snow was falling, making outdoor life very unpleasant. The two armies lay watching each other across the Rappahannock. Batteries of light artillery were stationed at intervals along the picket line. Captain Thompson's battery of the regular artillery occupied a position opposite the eastern outskirt of Fredericksburg. Thompson notified the general, commanding that he never omitted the custom of allowing his men unlimited whiskey on New Year's Eve and requested to be withdrawn from the front for that occasion. Being denied, he asked that a strong infantry guard be posted around his camp, as none of his men would be asked to do duty on that evening.

The 2nd Infantry was detailed for this duty, New Year's Eve and New Year's Day, while the men of the battery indulged in the free use of commissary whiskey. The relief on duty splashed their dreary round through the slush of snow and mud, while those off duty huddled close to a big campfire to keep warm. While we toasted one side and chilled the other, the bacchanalian revels waxed strong, and the sounds of ribald songs and boisterous mirth floated out to us on the heavy night air. On duty or off, the wet and cold prevented us from sleeping. During the night the "Grand Officer of the Day," Colonel Fenton, tarried awhile at our campfire. He told us the officers were having a "high old time" in camp and that considerable of the "creature" was afloat. The private soldiers had nothing to celebrate the advent of the New Year with, nothing to jubilate for, and no spirit for merry making. Discontent was very general. The men were dispirited and gloomy. There was a feeling that we had endured privations and hardships, fought hard battles, and squandered the blood of our bravest to gain ground, that had been lost and yielded to the enemy, through the incapacity of generals and the jealous disagreements of politicians, both in

Congress and in the field. The private soldiers felt that they were being used as tools for personal aggrandizement and were unwilling to be sacrificed for such causes.

This feeling, inactivity, and the discomforts of a winter camp began to tell on the discipline of the troops. Three weeks of inactivity dragged away. The absolutely necessary camp duties being all the men were called upon to do.

The view from our camp presented a dreary succession of camps planted in the mud. Fences and outbuildings had all been pulled down for fuel. The very few inhabitants that remained in their houses with intent to save their property were in a strait for provisions. They looked pinched with cold and hunger. Desolation and misery were theirs to the full. Respectable women became wantons from the direst necessity. Virtue was sacrificed for bread. History can never record the woes the private citizens of Virginia suffered. The "sacred soil" reaped a terrible crop from her secession seed.

[On January 1, 1863, President Lincoln issued the Emancipation Proclamation, which freed all slaves in territories held by Confederates and emphasized the enlistment of black soldiers in the Union Army.]

January 20 – Rumor was current that the Potomac army was to cut loose from its present base of supplies, turn the rebel left, subsist off the country and make its way to the Atlantic Ocean carrying the destruction through that part of the Confederacy.

January 21 – The weather had turned colder. The mud had frozen just enough to crust it over, but not enough to sustain man or beast. In a measure to confirm the rumors of yesterday, we had orders to be ready to move at any moment. Hooker's and Franklin's divisions began moving past our positions in the direction of Falmouth. The head of the column moved along in passable order considering the condition of the ground. But as the day advanced, the ground was punched deep into half-frozen mortar, and the artillery and baggage wagons cut deep into the soft road.

January 22 – A nasty mixture of rain and snow was falling and the still passing troops struggled heroically between the descending slush and the terrible roads. Despite their best efforts, they straggled badly. The artillery was plowing to the hub and their appearance was more that of a mob than an army.

January 23 – The dispirited troops began to return to their camps. Their artillery and baggage had been stuck in the mud. It was impossible to advance, so they returned.

The fagged out and gloomy stragglers from these troops were toiling past our camp until January 25. The most forlorn and pitiable objects imaginable. That the enemy had full knowledge of this move, and its outcome was evidenced by their displaying a huge banner on which was printed in large letters, "Burnside is stuck in the mud." This ended all attempts at a winter campaign, and the army settled down to brood over its woes while the talk of consolidating the regiments whose ranks had been decimated added a new factor of discontent.

The fame of his own regiment is dear to every true soldier and to have it obliterated by consolidation is unbearable.

The next three weeks was a season of general discontent. The soldiers were generally disheartened. Every paper that reached us from the north was full of abuse for this army because it did not annihilate the rebel force in its front, while we were nursing our unhappiness in our smoky huts, surrounded by an ocean of dreary mud. Discipline became very lax and desertions were very frequent. The only excitement we had was general review by General **[Orlando B.]** Wilcox once or twice a week. The men claimed that every time a congressman or a woman visited Wilcox, he got out in the mud to stand in companies for half a day that he might exhibit to them the size of his command. The smothered curses that were muttered against him on this account would have filled volumes.

February 10 – The 9th Corps received orders to be ready to move at a moment's notice. Leave not to return. This aroused a ripple of interest in the affairs of this world. Three days, we impatiently waited with knapsacks packed.

February 13 – At twelve o'clock noon, the order to march came. At Falmouth, we took the cars and were soon landed at Aquia Creek where we immediately embarked on the transport *Georgia*. Not knowing and little caring what our destination might be, there was a feeling of general relief at the prospect of leaving the mire and smoke of our winter camp behind.

Gloom, misery, discontent, and unhappiness were behind, the future did not concern now that the clean-decked ship floated us on the clear rippling water, with the blue sky overhead.

As we lay at the dock through the night, a feeble spark of hope seemed to take possession of the men that better days were coming.

February 14 – Casting off their moorings, the different vessels bearing the 9th Army Corps, steamed down the Potomac and out upon the Chesapeake. The bright sunlight glancing upon the waters, the fresh spring air, the passing ships, a school of porpoises at play, the flight of the gulls made today such a contrast to yesterday that life began to flow afresh in the soldiers' veins.

"Joy quickened his pulse, his hardships seemed o'er." **[From *The Inquisitor* by John Thomas Hope, 1822]**

As the beautiful day drew to a close, we sighted the long low projecting headland, Old Point Comfort. Sweeping around the point and through the channel between it and the Rip Rap, we came to anchor among the varied shipping in Hampton Roads under the guns of Fortress Monroe.

February 15 – Early in the morning, the anchor was up, and we crossed the Roads to Newport News Point at the mouth of the James River. Here we were told we should camp for a time to recruit health, restore discipline, and improve drill. The surroundings are admirably adapted for these purposes. The point is high and sandy, widening to a level plain of considerable extent inland, and the James rolls seven miles wide on the southern side.

The 9th Corps camped upon the level plain extending along the James. The position of the 2nd Infantry was about two miles from the point, facing the river. The plain at this point is about thirty feet above the river, and although the river is seven miles wide, the air is so clear that objects can be seen very plainly on pleasant days on the opposite shore.

Two gunboats of the *Monitor* pattern were anchored in the river to prevent the enemy from coming down with vessels.

From our camp the spars of the sunken *Cumberland* could be seen where she went down, colors flying and guns booming, carrying with her dead, wounded and able bodied, the living sending up a lusty hurrah as she plunged beneath the waves, sunken by the *Merrimac*.

Newport News was not a village, not even a hamlet. It was a mere landing, with a dwelling or two near.

Fish and oysters are taken in abundance from the brackish water near the mouth of the James. With the advent of the camp, of course, the oyster made a regular landing and numerous board shanties sprang up where cooked oysters and fish could be had for a small sum.

As soon as our camp was located, with the drum major I visited one of these shanties, presided over by an "Old Auntie," to get an oyster stew. Ordering a stew each, we were each soon served with a full quart of solid meats. Remarking to the "Old Auntie" that there was not much soup, she raised her hands and said, "Laud, love yer soul, honey! Oysters is what yer want – can dip up soup anywhere."

As we had come here for general improvement, the first care was to provide quarters that would be healthy.

New A tents were issued, and these were placed on frames four feet high, two tents together while our poncho tents were used to enclose the frames below the A tents. This made a roomy tent for eight. Material for these frames was gathered from all sources. Some of us went some miles up the river and tore down a barn, formed the lumber and timber into a raft and floated it to camp. Working in rain or sunshine, we soon had the most complete quarters that we had ever had since our enlistment. The encampment was laid out into streets and parades with exact precision, and as soon as the tents were arranged, strict orders were issued and enforced regarding cleanliness and order, and for a month, this became a camp of discipline and instruction.

Each day went through the following programs: six o'clock, Reveille; six-thirty, clean streets; seven, breakfast; eight, guard mounting; eight-thirty to nine-thirty, company drill; nine-thirty to eleven, battalion drill; twelve, dinner; two to four, brigade drill; six, dress parade; eight, Tattoo, eight-thirty, Taps.

Although we had a full share of rain, a healthy location, good shelter, plenty of good rations helped out with oysters and fish, plenty of hard drill and discipline, we grew vigorous and the old 9th Corps fast gained its former soldierly appearance, and in our busy life, we soon forgot the gloomy winter camp.

This camp leaves a pleasant memory, yet there were few incidents worthy of note.

A human skull was the football of our regimental parade ground during our stay at this camp. Whence it came, no one seemed to know. Where, when, or how its owner's life went out, no one seemed to care. Its shape would indicate that it was formerly worn by a Negro and on top the bone was more than a quarter of an inch thick. Kicked, hustled, and thrown about, it served as football, baseball, shuttlecock for the regiment and was never at rest only when the men were asleep. *"To what base uses we may return, Horatio. Why may not imagination trace the noble dust of Alexander till he find it stopping a bunghole."* **[From *Hamlet* by William Shakespeare]**

February 22 – A general review of the 9th Corps, with all the formal ceremony and parade of such an occasion was the incident of a day.

March 1 – February had slipped away, and March opened up without any change in our pleasant camp. Personally, I was given the rank of orderly sergeant, having performed the duties of that office for eight months with only the pay of sergeant.

Two weeks more sped on and our soldiers gained health and spirits. Constant drilling had corrected their bearing and general deportment. Confidence was fully restored, and the men felt that they were ready and willing to meet the enemy at any time or place.

March 18 – Orders came to have five days' meat cooked preparatory to a move.

March 19 – Turning our backs on our campground, we marched to Newport News at seven in the morning and immediately embarked on the steamer *Georgia*. About the time, we were on board, a storm of snow and wind set in, but we ran down to Fortress Monroe and anchored. The storm increased in fury, and for two days, we were storm stayed. Notwithstanding the shelter afforded by the roadstead, the water boiled and seethed around us like a whirlpool, and the sky was dark with wet snow.

To the common soldier, our destination was unknown. Neither did we know if our regiment alone or if the whole corps was to follow. Whether urgent or not, no ship could safely sail in such a storm, and we were compelled to wait until the storm abated.

March 21 – The storm was less furious, and the snow had changed to rain. The captain of the boat pleaded for another day, but it was not allowed him. Under compulsion, he ordered the anchor up, and at nine o'clock, we were under way. Heading straight up Chesapeake Bay, we ran a few points from head to the storm. The vessel rolled and pitched dreadfully, the hatches were closed, and everyone was ordered to stay below.

When we reached the horseshoe—the point where the current of the Potomac meets the swell from the ocean, the sea was exceedingly rough. Then officers, who had filled themselves with the sweets off the cabin table for two days, began to turn their thoughts from martial glory to their own true inwardness. Remarks about "New York" and "Europe" were loud and frequent and some indulged loudly in choice Russian language. For a truth, they were desperately seasick. Soon the close atmosphere was permeated with the smell of bile and fermented victuals, and the floors were covered with what looked like the unsightly contents of a swill barrel. In the stifling atmosphere of the steerage cabin below the water line, the privates were keeping down their hardtack, bacon, and unsweetened coffee very well. For me, the sights and sounds and smells began to cause a strange uneasiness in a moderately strong stomach. The feeling increased, and in a fit of desperation, I burst open one of the hatches and went on deck. The captain of the vessel met me and sternly commanded, "Go below, Sir." Recommending him to "Go below" himself, I walked to the bow of the boat. In crossing the huge swells, the bow of the vessel, in plunging down from the crest of one and rising to the crest of another, seemed to rise and fall thirty feet. As I stood for a brief time looking forward upon the surging waters, the uneasiness in my stomach took a freak impulse. A strong line seemed to extend downward to each of my toenails, while a mighty impulse seemed to urge the stomach upwards. Feeling the toenails start, I retreated to a safe place amidships and disregarding the advice of the capitan, I sat there in the pouring rain until we reached smoother waters, and my feelings were more composed.

March 22 – At an early hour in the morning, we arrived at Baltimore. The storm had passed, and the sun arose clear.

To prevent the men from straggling, they were kept on the boat until railroad transportation was ready.

The Ladies Aid Society sent an invite to the regiment to take breakfast with them, but for want of time, it was declined. A very different reception than we received in this same city less than two years ago. At nine o'clock, we disembarked and were loaded upon freight cars at Locust Point on the Baltimore and Ohio Railroad. For some reason, the train did not start until four o'clock p.m. Then our route lay westward along the Patapoo River, past relay house, Ellicotts Mills, Illcluster, and other pretty manufacturing towns on this clear rocky stream. The scenery was interesting as we sped away on this glorious Sabbath afternoon without a thought of our destination or its results. At ten o'clock at night, we halted at Harpers Ferry, got off the train, marched half an hour through the dark and muddy streets to get a slice of bread and a cup of miserable coffee, clambered back into our boxcars, and resumed our way.

Following the course of the Potomac through the mountains, we still moved westward.

March 23 – At Cumberland, Maryland, a brief halt was made for muddy coffee and a slice of bread, and the monotonous grinding of the wheels went on.

March 24 – A government breakfast was served at Grafton, West Virginia. From there, we took the Parkersburg branch. This road runs through a rugged mountainous region, and there are twenty-three tunnels. There was little of excitement on this jaunt in our inelegant cars; yet one little incident at Kanawha Station will show how quickly a body of soldiers can ruin a man's business. The train stopped at this station only long enough to take coal and water. There was a grocery and notion store adjoining the station platform. The men began to buy canned fruits, cakes, and other eatables, and offered bills on the broken bank of Erie and Kalamazoo. These being readily taken, everyone was anxious to buy. Such a rush for goods, the storekeeper never saw. Himself, family, and clerks were rushed to pass out the goods, giving out in change greenbacks, as far as he could, but if there was any trouble to get change, the accommodating soldiers would take any kind of goods for the whole money. The merchant remarked, "I never saw men as reckless with their money. They make trade lively."

As the train sped onward, after a five minutes halt, it bore away the entire contents of that store, and its keeper was left to count his gains in worthless bills on the Erie and Kalamazoo Railroad Bank.

Near sunset, we arrived at Parkersburg on the Ohio River, and were released from our three days confinement in cattle cars.

With little delay, we embarked on a new and elegant steamer, the *Majestic*, and as night closed in, we were sweeping down the Ohio. Spreading our blankets on the clean and roomy decks, we composed ourselves to rest, free from the rumbling cars and the cramped space and enjoyed a wholesome sleep as the boat held her quiet course between the uninteresting banks of the stream.

March 25 – Heretofore, we had attracted as little attention from people along the road as would a load of livestock. As we approached Maysville, Kentucky, a picturesquely situated town, there was a stir, and betimes, we were opposite the place, flags were flying, handkerchiefs were waving, and a cheering multitude thronged the riverbank. A like scene was enacted when passing Manchester, Ohio. Slowly it dawned upon us that we were out of the Confederacy, in the free North once more. So long in the heart of the Confederacy, we had been used to the contempt of civilians that these attentions started a new train of thought and aroused an interest in the scenes through which we were passing.

Late at night, Cincinnati was reached. Here we were asked to go ashore and eat, but the hour was so late, none was permitted to go. After taking on a quantity of coal, the course down the river was resumed.

March 26 – At twelve o'clock noon we landed at Louisville, Kentucky, over a thousand miles from our starting point.

The ladies of the Sanitary Commission were not aware of our coming until we arrived. They decided at once to give us a dinner at the Soldiers Home. We marched directly through the city to the Soldiers Home, and when we arrived, a ham sandwich and coffee were ready. Standing in column by section in the street, nice fresh sandwiches, cakes, and coffee was until all were filled. Dinner over, we marched out Broadway and camped on the common east of the city.

For months, we had been situated so that we had no commerce with civilians. Now that we were in a land of plenty, there was a danger of excess. Basket peddlers followed us to camp and before the tents were up, were busy selling pies, cakes, and such fruits as the early market afforded. There were scores of Irish and German women with well-filled baskets of things to tempt a soldier's appetite.

The strictest guard was kept over the camp, and the rules against liquors were enforced, yet it was soon evident that more than one man was the worse for liquor. A little search revealed the fact that most of the baskets had bottles of whiskey under the cakes. Then the baskets were searched before they were admitted to camp. But the men kept on getting drunk just the same. A little detective work showed that the women were taking the men's canteens out and bringing them back under their skirts full of whiskey. Then the peddlers were entirely excluded from camp, but the drunkenness did not stop. Where the supply came from was a mystery and remained so during the two days of our stay at Louisville. It subsequently leaked out that men who went out of camp under arms, by plugging the nipple of their muskets with a bit of soft wood, could put a pint of whiskey in the barrel, stop the muzzle with tampon and safely bring it into camp. Many gallons were undoubtedly smuggled into camp in that way. As we stayed but two days near the city, these temptations were soon out of reach.

March 28 – As the sun sank low in the west, we found ourselves on board the cars on the Louisville and Nashville Railroad bound out of Louisville. Our departure from Louisville was without incident save that two officers were detailed to search the train for whiskey. In the careful discharge of this duty, they smelt and emptied all the canteens of water that the men had slung to them, and left two full canteens of whiskey that hung on hooks in plain sight, thinking surely that the contraband stuff would not be exposed so boldly. Truly, the tricks and arts are past finding out.

March 29 – Early dawn found us at Bardstown, Nelson County, Kentucky. Before the sun was up the tramp of our column roused the peaceful slumbers of this quiet village, as we moved to a camping place on the plantation of ex-Governor Wickliffe.

This town, though not on very high ground, is undermined with a large cave, which the inhabitants utilized during the frequent raids of Morgan to hide their movable property. The surrounding country is fertile and lovely, and the people more intelligent that those we met in Virginia.

Seemingly, our brigade was sent into this state to check the movements of John Morgan and Champ Ferguson, the noted guerilla chiefs. Our little brigade being the only Union forces in this vicinity, the utmost vigilance was necessary to prevent surprises by these raiders. This fact furnished plenty of active duty, although there was no enemy in sight.

March 31 – My company was on picket and a heavy fall of wet snow made it very disagreeable for us.

During our brief stay at Bardstown, nothing left so great an impression on my mind as a pair of shoes that would correspond for the regular size of twenty-one, which the dealer said was made for a Negro of natural proportions.

CHAPTER 11 – APRIL 2, 1863

We left Bardstown by the Lebanon Turnpike to begin a tramp of some fatigue, but of little danger, past the green fields and beautiful scenery of central Kentucky.

At night, a halt was made at Springfield, and the fairground through which a clear pebbly stream made its way, afforded a fine camp and a good place to bathe our blistered feet.

April 3 – Resuming the march on the Lebanon "pike," the variegated scenery of a delightful country spread out on every hand, enlivened by people engaged in rural occupations, being a strong contrast to the barren and fenceless fields of the tenantless plantations in Virginia. For two years, we had not looked on much else than warfare and its necessary adjuncts.

For some weeks, we were now to be a "Speck of War" amid peaceful life and be surrounded by civil social life, of which we were no part.

Passing through the town, camp was pitched a mile east of Lebanon, on rolling ground near a fine creek. The strictest discipline was enforced, both to prevent surprise from raiders and to prevent stragglers from isolated brigade from committing depredations on the inhabitants. Drilling and their military duties filled up the time pretty well, but there was an occasional incident and some opportunity to study the manners and customs of the people.

Just at this time, I received a warrant as orderly sergeant, the duties of which office kept me in camp through the day, but was permitted to go to the town for a couple of hours after sundown to get a "square supper."

Being recommended to a Mrs. Custin who kept a boarding house, I took tea there with great regularity for several days. That lady was of strong secession principles, but most of those who sat round her table were Unionists. The good lady strongly objected to the Yankee sound of e-u-w in cow and like "Yankee drawl," but when informed that a native of her own town had been heard that day to say, "I went to C-y-arter's to borrow his c-y-art to haul some c-y-abbage to town," she pleasantly admitted that the Southern drawl was no better. At another time, a doctor's wife brought to the table a turnover pie she had bought of a huckster woman who hawked pies among the soldiers. Mrs. Carter chided her for buying such stuff when her own pies were so much nicer, and she could have as much as she desired. The lady informed her that she wanted no pie at all, but was desirous of knowing what sort of pie soldiers had to eat. Then she proceeded to dissect and taste her purchase with many grimaces and witty remarks and by declaring she believed the crusts were made from the dried cast-off of the beeves slaughtered for the army and filled with dried toadstools from logs in the woods.

April 10 – A brigade of troops under General **[Mahlon Dickerson]** Manson, in which were the 18th and 22nd Michigan regiments, occupied Lebanon. In the 18th was a brother and many of my old friends with whom I spent an enjoyable day.

General Manson, being the senior officer in that neighborhood, assumed command of the whole country and issued orders that no soldier would be allowed in the streets of Lebanon without his pass. About noon, I saw a soldier running from the direction of the town pursued by two soldiers with arms in command of a man without hat or coat. The fugitive soldier proved to be one of my company and dashed into camp closely followed by the others. As he rushed into his own tent, the hatless and coatless man shouted, "There he is. Take him."

Not liking this kind of maneuver, I quietly ordered five or six of our men who were standing by to "Fall in with arms." Then approaching the intruders, I inquired, "Who are you, and what is your trouble?"

"I am General Manson's clerk, and these are guards from his headquarters. He was in Lebanon without a pass, and I was ordered to arrest him by General Manson."

"You cannot take a man out of this camp in such a guise, if you are an officer on General Manson's staff, and nobody's clerk can take a man at all. You get out of this camp in two minutes or I will take you and your guard to the colonel."

"That is just the man I want to see."

"Very well. Men, take these prisoners in charge."

Conducting them to the colonel, he sent them to the guardhouse, but kindly allowed them to send a message to General Manson as to where they were. General Manson was compelled to send an officer to procure their release, and they departed less confident but wiser than when they came. The idea seemed very ludicrous that a clerk should attempt to arrest a soldier.

April 11 – Colonel O.M. Poe, who had led us for most of two years, called out the regiment and informed us that he was no longer our colonel, and was about to leave us to resume his place in the Engineer Corps of the regular army. Many desired to go to the depot to bid him and his wife good-bye.

Captain Humphrey, upon whom the command of the regiment devolved, issued orders that, all who wished, could go under the command of the orderly sergeants of the various companies. Soon the entire enlisted force of the 2nd was formed in line upon the regimental parade with white gloves, and side arms. On consultation, it was decided that, my company being on the right, I should execute such maneuvers as I pleased, and all others would execute the same movement.

The commissioned officers of the 2nd had the rare sight of seeing the enlisted men of their regiment, without an officer, parade through the streets of Lebanon, executing with mechanical precision, "On the right – front into line," "On the right, form column, "By the right of companies, form column," "Right, oblique," and various other maneuvers more steadily than when under their command.

Arriving at the depot, ranks were broken and all shook hands with our old colonel, and when the train that bore them away sped out of the depot, Colonel Poe and wife were sped on their way by three rousing yells for Colonel Poe and wife.

The train being gone, ranks were reformed and the return march was made without an act to mar the reputation of any of the participants.

This section of Kentucky is devoted to the raising of grain, tobacco, cotton, and fine stock. The only thing to mar the generally level surface is the abrupt sugarloaf-shaped hills, called "Knobs" by the Kentuckians that rise up here and there over a large section. About these knobs, many fine fossil specimens of snakes, fish, shells, and other things may be found. Also, a boulder is plentiful that when broken, the inside is hollow and thickly set with many-sided crystals. In my short strolls, I had gathered quite a collection of these, but when we left Lebanon, they had to remain behind.

April 27 – Lebanon was left by the Columbia Turnpike, and camp was made by Campbellsville, twenty miles to the south, and the next day a march of ten miles brought the brigade to the Sublett farm on the banks of the clear, rocky-bottomed Green River. The 20th Michigan was here detached and sent farther south toward the Cumberland Mountains, while the others were left here to guard the country and rebuild the turnpike bridge that Morgan in one of his raids had burned. The heavy timberlands along Green River afforded ample supplies for bridge building, and as there were carpenters more than enough to build the bridge, the majority were left for several days with ordinary camp duties, drills, and an occasional turn on picket to pass the time in this peaceful and picturesque country. The idle monotony of our existence was broken only by an occasional incident or a stroll across the country.

Most of the people were loyal to the Union, but in some instances, there were found disloyal people. In either case, the feeling was intensely bitter. An incident will illustrate this feeling and one of our amusements at once.

May 2 – Corporal Curtis and myself devoted the warm and pleasant day to a ramble along the river, between its clear waters and the overhanging cliffs, near the mouth of Robison's Creek, in search of anything good to eat or any scraps of information that might fall in our way. As noon approached, we cast about for a place to dine. Lighting upon a farmhouse that looked clean and neat, our request for dinner was granted, if we could wait while it was cooked. While the meal was in preparation, the corporal was using every art to coax the children to him with no success, and the matronly woman of the house was giving me a detailed account of her nearest neighbors who were "Pizen Rebels," she said.

The Ronynes had a son in the Confederate army, whom she said, had lately been seen skulking round house, and I was exhorted to capture or kill him, which I promised to give my attention. While I had been giving my whole attention to the mother's talk, wholly indifferent to the children, they had, one by one, shied around the corporal and taken refuge in my lap, four in all. The corporal had offered them everything his pockets contained, to come to him, without avail, and now they were offering me little presents from their few playthings. The corporal good-naturedly said he would get ahead of me before we reached camp.

Striking northward from the farmhouse about two miles, we next called at the house of Mack Steiger. The daughter of Steiger took upon herself to extend to us a somewhat superfluous welcome and also did nearly all the entertaining. She was a spontaneous talker and not altogether uninteresting.

The corporal, without waiting to learn the young lady's war sentiments, boldly expressed disgust with the war, and declared himself heartily tired of the army. On the other hand, my voice was for a vigorous continuation of the war with a willingness to spill my own blood as free as water. After listening to our war views, briefly stated, the young lady at some length expressed her great admiration for the fine defenders of the country, ending with the statement that if she had a husband, she should want him to be in arms against the rebellion, and that she had no sympathy with those who would not sacrifice love, happiness, and life for their country.

The corporal thought it time to return to camp, and on our way thither, he declared in all good humor that he had no further use for me as a fellow stroller, as I seemed to be his rival on all occasions.

The next day being Sunday, several teachers with their pupils, witnessed dress parade at our camp, these being the only troops that had camped in this vicinity. Great pleasure was expressed at the spectacle, and they departed thinking they had beheld a wonderful military exhibition.

One pleasant morning, a Jew peddler with a pack filled with watches and jewelry, appeared at camp. With great expectation of gain, he displayed his watches to the soldiers who crowded around, as he dwelt upon the superior excellence of his stock. Handing out a dozen or more watches for examination, they passed from hand to hand as one after another critically scrutinized the workmanship and movement. He soon missed some of the watches and investigation showed that several were missing, and no amount of search could discover them. Closing his pack, he went to Colonel Morrison, who was in command of the brigade, demanding payment for the watches his men had stolen. Colonel Morrison, casting a fierce look upon the Jew, thundered, "My men steal! Get out of this camp in five minutes, or I'll have you arrested as a spy." The Jew left.

May 7 – Rumors reached us that Hooker had won a great battle at Chancellorsville. Much excitement prevailed, yet the older soldiers wanted to hear more definite news before they would make a demonstration.

May 8 – News of Hooker's defeat deadened the rejoicing over the first report.

[The Battle of Chancellorsville, April 30-May 6, 1863, resulted in a Confederate victory, with 30,000 men killed or wounded.]

May 10 – Picketing north of Camp Sublett, our line lay along an angling position of Columbia pike, and extended across the fields in continuation of the pike at either angle.

It being Sunday, many people passed on their way to a little church farther north. A couple of young ladies returning from church with a horse and buggy dropped a handkerchief near the first picket post. The soldier on duty halted the buggy, returned the handkerchief, and allowed them to proceed. At the next post, the handkerchief was dropped and returned as before. Having witnessed the other two performances, I was surprised to see it dropped opposite my post. Halting the carriage, I raised the handkerchief and told the ladies it was a pity their two former offers had been refused. I would not refuse the handkerchief, but would keep it in remembrance of their generosity. Blushing scarlet, the young ladies put whip to horse and rode away. The two giddy girls were strong secessionists and had adopted this frivolous method of making the despised Yankees wait on them. Southern ladies were not above such trifling.

May 11 – By orders, the picket line was abandoned and upon return to camp, we found it deserted. The 20th and 17th had encountered John Morgan **[Brigadier General John Hunt Morgan known as the "Thunderbolt of the Confederacy"]** near Monticello, and the other regiments had hastened to their relief, leaving us to follow.

Following the turnpike southward thirteen miles, the 2nd, 8th, and 79th were found in bivouac at Columbia. The next day, the 17th and 20th came in and the troops were disposed on high ground in good position to an attack should Morgan think it advisable to fight more. After waiting a couple of days, learning that the enemy had retreated, a convenient place for camp was sought in that vicinity.

May 14 – Camp was located south of Columbia near the Jamestown Road on Russell's farm. The location was a delightful and healthy one. The country was divided into small farms that were under cultivation with also much heavy timber. Company D went on picket along the Jamestown Road to the south in a piece of heavy thick woods. Lieutenant Van Buskrik and myself were sent to explore the woods for two miles to the east while two others went to the west in search of any byroads by which an enemy could rapidly approach. Nearly a fortnight passed in quiet at Camp Russell's farm.

May 27 – A military officer inspected our camp and pronounced it healthy. Late in the day came an order to the 2nd to be ready to march at seven in the evening. Promptly at the hour, the march southward began and continued over rocky road, through a broken and hilly country nearly all night. At seven in the morning, the march was resumed, and Jamestown was reached at noon.

Bivouacking near the town, scouts were sent out in various directions. Toward night, a heavy rainstorm set in which continued through the night.

May 29 – Some men who were out foraging were fired upon by guerillas.

Company I was sent out to reconnoiter and brought in a prisoner with a splendid chestnut horse he had stolen from a neighboring plantation. Our chaplain, who was much better versed in horse flesh than in scripture, looked the animal over critically and offered the colonel one hundred dollars for him. Scorn and contempt mingled in the countenance of the colonel as he replied, "Chaplain, that horse is not mine."

Afternoon, Company D was detailed to scour a thick matted wood east of the town for guerillas and horse thieves. The rain was falling like a torrent, and the undergrowth was so thick that it had to be parted with the hands to make a passage. For some hours, we beat the bush, receiving the deluge of water from above and having it rubbed in by the mass of bushes and running vines, until with clothing and flesh torn and water soaked and chilled, the end of the wood was gained, and no game had been started. Returning to the bivouac, which was at the corner of an open wood, an immense long heap was built and fired.

The rain ceased at nightfall, and the men ranged around the great fire to warm their benumbed bodies and dry their clothing. The intense heat acting upon the wet garments soon treated each to a hot steam bath, which probably prevented much bodily harm. Turning like a spit before the fire, till near morning, the men became dry enough to get a brief sleep.

May 30 – Orders were received to return to Columbia. Although it rained and was very muddy, the 8th set up a banter for a trial of endurance on the march, which was accepted by the 2nd, and the march began and was maintained with great spirit all the way. The result was that the 8th came straggling in at the end without any military formation, while the 2nd, with ranks in order, had only fifteen men absent from the ranks.

June 4 – After a day or two of notice, having sent off all extra baggage, the whole brigade returned to Lebanon by the way of Camp Sublett and Campbellsville, arriving at Lebanon at noon the next day. Here, four days rations of meat were cooked, and we took the cars for Louisville at sunset.

The trainmen were uneasy from fear of trouble with guerillas, a camp of whom was said to be located four miles from the line of the road. Louisville was reached at 3 a.m. without any trouble, and we immediately crossed to Jeffersonville, Indiana. There seemed to be a noticeable difference in the feeling of people toward the soldiers as soon as the Ohio River was crossed.

The open suspicion, or at least cold indifference of the southern people, was replaced by expressions of confidence and friendship by the Indiana people. At three p.m., we left Jeffersonville by the Indianapolis and Jeffersonville Railroad and were soon whirling through the little towns along the line, saluted everywhere by the waving of handkerchiefs and patriotic cheers by the ladies and children. At Seymour, we took supper and changed cars to the Ohio and Mississippi Railroad, bound west.

June 7 – Breakfast was served at Vincennes, a neat and thriving city on the Wabash River. Still moving westward, our progress was marked everywhere by patriotic demonstrations. At Clay City, Indiana, and other places at which we stopped, the ladies met us with baskets of provisions, fruits, and bouquets, giving words of good cheer and encouragement and some did not hesitate to bestow kisses. Being so long accustomed to the indifferent manner of the people among whom they had been, the hearts of the soldiers were cheered by the interest exhibited by these loyal Hoosiers. Who shall say how many were braced up to heroic deeds by the encouragements given on this trip through the loyal north? At Sandoval, Illinois, cars were again changed to the Illinois Central, bound south. Supper was taken at Centralia, Illinois, a fine place, surrounded by broad stretches of prairie, dotted all over with farm houses.

As the train sped southward, the descending sun gleamed across the broad expanse of gently undulating prairies, clothed in the fresh verdure of early June, tingeing the tops of the hedgerows with gold and glancing its beams from the farmhouse windows in spikes of flame. The breath of early summer was in the air, and the corn was limitless. From close at hand, away to where the earth and sky meet, houses, hamlets, and villages could be seen with their orchards and cattle, lending an added charm of domestic life to the natural beauty of the scene. Before darkness shut out the view, a town was passed that was located on a gentle rise in the otherwise level prairies. The name, Richview, aptly describes the scene as we sped past, going we knew not whither.

June 8 – Daylight found us at Cairo, situated at the confluence of the Ohio and Mississippi rivers. The town and surrounding country is on a level with the river at the ordinary stage. A heavy dyke protects the country from an overflow. The business places were in a line along the levee on the Ohio River, the levee being the street. Many of the smaller houses were built on flat boats and in some cases, the sidewalks were as high as the top of the first story of the houses. Heavy rains would fill the streets several feet deep with water, and a large steam pump was in position to pump the water out of the town over the levee. Altogether, it was not a desirable place of residence, but as a shipping point, it was of considerable importance and was of great use to the government as such.

Toward the evening, the 2nd embarked on the Steamer Nebraska, but remained moored to the dock all night. Steamboats on the western rivers differ from those on the lakes in that the hull is broad with the bottom flat, the largest boats drawing only three and a half to four feet of water. The boilers and engines are on the lower deck instead of in the hold, and the great furnaces are at the open bow of the vessel. The hold is very shallow and holds but very little freight. After the hold is full, the loading is piled on the lower deck at the bow and along side the boilers, and in fact, all over the vessel.

June 9 – Early, the Nebraska cast off her mooring and steamed down the Mississippi, properly speaking, the lower Mississippi. This great river is held in bounds by low, marshy, monotonous banks covered mostly by cottonwoods, except where an occasional bluff extends to its side, and upon every bluff, there is a town. The first of these is Columbus, Kentucky.

At that time, it was a military post, and our boat touched there for final instructions. It was then given out that we were to join Grant at Vicksburg. So the Nebraska was to be our home for a voyage of six hundred miles down the river from Cairo.

There was a large yawl boat towing astern and for this, four of us left the crowded decks, taking our passage there, only going aboard for our rations or to take a turn on guard, although the captain of the boat assured us, we would surely be upset and drowned. The hardihood of old soldiers would not allow us to be frightened from our position, and we rode in the yawl to the end without mishap.

The next break in the general monotony is Island No. 10, so called because the islands of the Mississippi are all numbered on the charts in regular numerical rotation. This island is famous for the length of time the rebels held it against all efforts of our people to take it. The reason of its strength may be seen from a description of its situation.

Approached from above, it appears like a square block of sand, rising twenty or thirty feet abruptly out of a low marsh, just even with the river.

The only feasible approach to this island is by water, and as it is approached from above, the river forms two long loops in the low marsh, so that a vessel would have to travel four miles, all the time being not more than one mile from the guns on the island, and then the channel runs straight past the island within a stone's throw, makes another curve at the lower end and passes on its way. A hostile vessel would be long enough under fire to be knocked into kindling before it would get to the island.

After passing New Madrid and Mount Pleasant, Missouri, night came on, and as it was unsafe to run after night on account of guerillas, the boat was tied to trees on the Tennessee shore, and pickets thrown ashore while those on board slept till morning.

June 10 – With the light of day, the Nebraska was moving down the river, the unchanging scenery broken only by Fort Wright and Fort Pillow until four o'clock p.m., when Memphis, Tennessee, was reached. This city is of some note, but trade had been stagnated by the war, and most business houses were closed and tenantless. The streets and alleys were in a wretched condition for want of repairs and from accumulated filth.

The next morning, the steamer ran over to the Arkansas side, and the troops disembarked while the vessel was being cleaned, and then re-embarked and returned to the levee at Memphis for the night.

June 12 – The trip was resumed. Helena, Arkansas, was passed early in the day. The town stands on a level plain, elevated a few feet above the river, backed by quite considerable mountains that seem to overhang the place. The whole scene is picturesque, and the place is of importance in a military point of view. At night, the boat was tied to the Arkansas shore. Stepping ashore and moving back into the dark woods a little, the scene on the Nebraska seemed almost supernatural. Of the greater portion of the vessel, but a dim outline could be seen. Around the lower deck at the bow were placed iron jacks holding blazing torches. Eight Negroes, stripped to the waist and streaming with sweat, shoveled coal into the sixteen fireplaces under the immense boilers. The glare of the torches, the opening and shutting of the fiery furnaces, the perspiration on the bodies of the Negroes shining like metal, backed by the weird outline of the steamer and the black darkness of the night around suggested thoughts of Dante's Inferno.

June 13 – Passed Napoleon at the mouth of the Arkansas before breakfast. It is hard to tell whether this place is above or below the mouth of the Arkansas. The alluvial soil is so loose that at high water, the river chooses its own channel and has run both above and below the place, leaving it virtually an island.

Soon after passing Napoleon, while we were squatted about the decks eating our breakfast of coffee, hard bread, and corned beef, a rattle of musket fire greeted us from guerillas from behind the levee. Instantly, the coffee was set down, and guns were taken, and a murderous volley was poured upon the top of the levee from all three of the decks. Such a concentration of fire raised a wonderful dust, but we knew not if any rebels were hit for we saw none nor heard more of them. The casualties on the boat were two men of the 20th Michigan and a mule slightly wounded.

Later in the day, Lake Providence, Louisiana, was passed. The lake itself is a basin two or three miles in extent each way, having an inlet from the Mississippi, and is surrounded by a sandy plateau a few feet above the river.

Night brought us to Milliken's Bend, Louisiana, where the Nebraska was moored for the night. We were now within sound of the guns of Vicksburg and were practically a part of the besiegers, although a day or two might be necessary to find our field of operations.

[The Battle of Vicksburg went on from May 18 to July 4, 1863.]

CHAPTER 12 – JUNE 14, 1863

The *Nebraska* was run down to Young's Point, and the troops disembarked on the Louisiana shore near the famous Vicksburg cut-off and in full view of Vicksburg.

The town stands on the high bluff on the Mississippi shore, where the river curves sharply toward the bluff, runs just past the town, and then curves away again, leaving a long low point on the Louisiana shore opposite the village. At the base of this point, Grant started the famous "Vicksburg cut-off." The ditch, or cut-off, extended across the low land from above Vicksburg to intersect the river below, leaving the town a mile or two inland. Had sufficient dirt been excavated to let the water through this channel, Vicksburg would have been practically an inland town. But there had not been enough dirt thrown out to convince us that Grant ever meant to leave the river flow through his ditch. Private soldiers expressed the opinion that this was only a subterfuge to gain time, keep the men busy, and fool Congress until he was ready for a better move. The 9th Corps had come over a thousand miles to assist Grant in capturing Vicksburg, and we being in advance were held here on Young's Point two days until the troops all arrived. During this two days, there was opportunity to observe the surroundings.

Along the low shores, at intervals, were low, broad and flat boats of very stout build, each supplied with a mortar of heavy caliber, from which every half hour, ten and fifteen inch shells were projected high into the air, curved gracefully toward Vicksburg, dropped into the town and exploded. From the number of mortar boats, the occasional reply of the enemy and the bursting of shells, there was a lazy kind of cannonading going on all of the time. At the burning fuses of the shells would mark their lines of flight and their graceful curves in various directions produced as fine an effect as holiday fireworks. Along the cut-off, a line of sentinels was posted to keep the soldiers from getting within range of the enemies guns. This duty was performed by Negro troops, the first we had met, and it was hard for some of the boys who desired to push their investigation farther toward Vicksburg to be stopped by a "Damnigger."

It was often illustrated that the more ignorant a soldier, the more stringent he would be in obeying the immaterial parts of his instructions. If these colored troops had been more familiar with firearms, several of our men would have been killed or maimed. The trespass of a foot across their beat was sure to bring a shot from their old smooth bore muskets. But as they had not learned the use of sights on a gun, someone else was more liable to get hit than the man who had ventured too far.

The bottomlands extend miles back from the river and are wooded back of the cut-off. In this timber, was a picture of misery characteristic of the times following the emancipation proclamation. Fifteen thousand Negro refugees—contrabands—were assembled—it could not be said they camped. Without shelter, food, or sufficient raiment. At the first note of freedom, they had flocked in from the back plantations without any provision for the future. Sprawled about under the trees on the bare ground, without covering and half-clad, were the aged and gray patriarch of the plantation, the old granny, the Aunty, the lusty buck, the buxom wench, and the little piccanniny, promiscuously tossing their limbs about in fight with mosquitoes and gall nippers, while the intense heat, only partially subdued by the shade of the trees, kept the sweat pouring from their bodies like rain. Some, from inhaling the malaria of the low ground, exposure to the sun and storm and want of food, were sick with fever. The most aristocratic of them could boast of no more than an old piece of tent cloth, tied by the four corners to the trees for a home, or an old quilt to lie on. Most of the children were clothed only in a sleeveless shirt, no covering for the head, anus, or legs. The fume that rose from this feverish and sweating mass of black humanity was as stifling as the hold of a slave ship. The government distributed bread and meat to these fifteen thousand souls for weeks, but there was no means of shelter at hand. They had been raised from childhood to be provided for by the master, and they had yet to learn to think and plan for themselves. They were as little capable of taking care of themselves as a child of five years.

It had been expected we would cross the river below Vicksburg at Warrenton, but after having moved to a point opposite that place, we were ordered back, traversing the cut-off its entire length to the place where first we landed.

June 16 – The steamer *Kentucky* took us on board and ran up the Yazoo River. This river empties into the Mississippi a little above Vicksburg, skirting the bluffs to the rear of that place. We were warned by the steamer's company not to drink or bathe in the water of this river, whose name means death, and whose waters are as sluggish as death. As we lay moored to the bank at Snyder's Bluff, a block of wood thrown into the water at night had not perceptibly moved in the morning. Drinking or bathing in the slimy green water is said to produce an aggravated kind of malaria that produces death in a few hours. The Yazoo is navigable some distance inland and connects with numerous navigable bayous. Cypress swamps filled with reptiles are a feature of its headwaters. On the south side near its mouth are high bluffs, the more notable of which are Snyder's and Haynes bluffs that receded steeply from the water to a considerable height.

June 17 – We disembarked and moved up onto the bluff and inland about two miles to Milldale and from Vicksburg about twelve miles. The firing at Vicksburg could be plainly heard, but the newly arrived were not to be used in the siege direct.

The 9th Corps with other troops under General Sherman were set to watch the rebel General Johnston who was threatening Grant in the rear.

With our backs to Vicksburg, fortifications were constructed and other precautions were taken to hold Johnston back from giving aid to the besieged Pemberton. The earth of the bluffs is a dry red sand, and they are furrowed with wide and deep ravines, rendering the passage of teams impossible except on the regular roads. Water is scarce and wells are unknown in a large tract of country. The inhabitants universally use cistern water for all purposes. A creek furnished our camp with water. As there were several days without rain, our creek became pools of soapsuds. The nearest supply other than this was a flowing spring two miles distant. My "pard" and myself made a trip to the spring once a day each with our two canteens, and when there, we fell in line with fifty to a hundred others and waited our turn at the spring. When at the spring, no one was allowed to fill more than one canteen at a time. Once filled, we dropped to the foot of the line and patiently waited our turn to fill the second. This supply—six quarts a day—had to do for drink, washing, and cooking purposes for two of us.

Thus the days dragged on—Grant pounding away at Vicksburg and Sherman taking care of Johnston, with an occasional incident of the siege drifting back to us.

June 20 – Pemberton, to save the forage his useless mules consumed, and thinking that the Yankees would be greedy to capture them, gathered several hundred mules, stampeded them through Grant's lines, and attempted to follow with his troops. The "Yanks" opened ranks and let the mules out, but closed immediately the gap and kept the rebels in, repulsing them with great slaughter. With fighting within hearing all the time and plenty of military duty to do in our own field of operations, we still got a day off sometimes.

June 22 was one of the days off for me. Leaving the bluffs with some comrades, we sought the Yazoo flats in search of blackberries, which we found luscious and plentiful. Having filled our little tin pails with berries and eaten all we desired, we visited a little house where lived an English schoolmaster and wife with several small children. Obtaining a drink of water, we were about to depart when the good lady proposed to take our berries and make us a pie. Now, pie was what no soldier could resist, so we pooled our berries and fifty cents in money for a pie. Betaking ourselves to the shady side of the house to wait its advent. The children found me very attractive and were soon searching my pockets for "pretties" as they termed it. Finding several photographs, these had to be shown to the mother. She was inquisitive to know who each was. When I claimed them all as sisters, she wondered at the size of our family, but insisted there must be another one who was my "lady love."

Just at this time, it was announced that the pie was ready. Drawing around the table, a milk pan with what seemed to be a steaming loaf of bread in it was placed in the middle. Experiments with knives proved it to be pie. A gutta-percha sort of crust had been placed in the pan, the interior filled with berries and water and a crust lay over the top. This was baked without sugar. With considerable twisting, the hard portion of the pie was disposed of, and we departed some wiser, if not highly pleased.

An occasional incident like this was all that broke the dull monotony of the firing at Vicksburg and the ceaseless round of military duty.

June 29 – our camp was moved to Flower Hill Church, six miles nearer the Big Black River, which was the dividing line between our forces and Johnston's rebel army.

Flower Hill Church is a modest little chapel in the woods at the crossing of two roads. Water for cooking and drinking purposes was very scarce. Those who desired to wash their underclothing were compelled to go several miles for water. An attempt to dig a well in the bottom of one of the deepest ravines revealed nothing but dry red sand as far down as the digging went. The oldest inhabitant never heard of anyone digging for water and finding it in this section. All water that falls upon the roofs is conducted through filters into cisterns for all uses. As in camp at Milldale, the same dull round of camp, picket, and fatigue duty prevailed, with provisions a little scant. Peaches were ripening and blackberries were plentiful. These helped to keep back hunger to a great extent.

July 4, 1863 – At eleven o'clock a.m., it was announced that Pemberton had surrendered unconditionally to Grant, and that the siege of Vicksburg was at an end. Three hundred pieces of artillery, 27,000 stands of small arms and 32,000 prisoners were the estimate of property fallen to us and another long stretch of the Mississippi was open. Enthusiasm ran high in our army at this glorious news.

[Records revel 19,000 soldiers in all were killed or wounded. Meanwhile, the Battle of Gettysburg occurred July 1-3, bringing

another Union victory. Total killed or wounded: 46,000.]

At four o'clock p.m., the portion of our army under Sherman was put in motion against Johnston. Late at night, we were moved by the flank from the road into the pitch-dark woods and ordered to bivouac. The inky darkness prevented any elaborate bed making and each man dropped down where he was and went to sleep. Thoughts of insects and creeping things never disturbed the dreams of a weary soldier.

July 5 – Although it was a glorious Sabbath morning, bivouac was broken at two in the morning. To prevent surprises, the country was swept thoroughly with troops as we advanced. The line of march assigned to the 2nd led through a large cypress swamp. It being the dry season, the swamp was passable, but the stubs of cypress trees showed us where they had rotted off between wind and water in the wet season, ten feet above our heads. Snakes and lizards whisked away at our approach and alligators were looked for but none was seen. The cypress trees with their straight stems rising fifty to seventy feet without a limb and then around a beautiful top, were a marvel rising out of their revolting surroundings. The profusion of gray moss was the only other redeeming feature of this dismal, hot stifling slum. It hung dead branches like long gray whiskers. It spread itself over slumps of bushes, forming fantastic bowers for reptiles. It climbed the tall cypress trees and hung in silver festoons from the top.

Out of this disagreeable swamp, we emerged late in the afternoon and camped in a splendid grove where magnolia and fig trees abounded. The dark leaves of the magnolia, at a little distance, had a warm satin luster, but on a near approach, the appearance becomes cold and reserved. From a little distance, the blossom and foliage are luscious and inviting, but on nearer inspection it seems to repulse by its frigid dignity.

Resting here for a day in this dark wood with clouds lowering overhead, news came that Johnston had retreated to Jackson.

July 7 – The camp was aroused at 4 a.m. with orders to be ready to move at five, but as was frequently the case, the men were kept waiting with cartridge boxes, knapsacks, canteens and haversacks at low until 2 p.m., before a move was made.

The Big Black River was reached and crossed about four o'clock and the advance was slow and cautious.

One of the sudden thunderstorms peculiar to the section came on; the heavens darkened to midnight—darkness and sheets of water came down with such fury that the men were powerless to make head, and they stood amid the deluge for a time, helpless and unable to see, except by the blinding glare of lightning flashes that swept the sky, leaving the darkness more intense by contrast. The bellow and roar and growl and rumble of thunder was incessant, crash following crash without intermission. For an hour, the march was effectually stayed. Then the cloud rolled on, and the sun came out for one parting glance before it sank to rest behind the low swampy woodlands along the Big Black that we had left in our rear. Half-blinded, nearly stunned and entirely drowned, we moved on. Night came on, the sky was again overcast, and darkness again shut out everything from sight. Plodding on, lifting great clods of clay at every step, gaining a little confidence from an occasional flash of lightning, drenched, exhausted, and hungry, we at last camped eight o'clock near the Jackson and Vicksburg Road. Wet as everything was, it was a task to light fires. We knew by past experience that to find wholesome water in this section was impossible in the darkness. The little rills running in the gutters that had been filled by the recent storm furnished water, thick enough with clay for gruel, to quench our thirst and make our coffee. No need for milk, no use in the world for cream in coffee produced from such a mixture.

July 8 – Morning dawned cloudless, and the sun shone out bright and beautiful, yet the light of day revealed to us a broad sweep of wet clay, which we worked into mortar wherever we trod, and gradually plastered our clothes with it as we moved about. The distress for water was very great and rations were entirely gone. The quartermaster was sent out for anything to eat. About ten o'clock, he came in, his guard of armed men driving five little, poor, starved yearling calves. As it chanced, he drove them close by the bivouac of the 2nd Michigan. Disregarding the authority of the quartermaster, and caring nothing for the guard, the men stepped out, knives in hand, laid hold of one of the yearlings, cut its throat and each skinned such a space as would give him the cut he wanted, sliced off a piece and had it over the fire without waiting for the death throes of the beast or the animal heat to cool. Nothing ever stands between a hungry army and food, except the entire absence of anything eatable.

The move toward Jackson was resumed at one o'clock. The heat that had been temporarily cooled by the rain began to be oppressive again, and no water could be found except in cisterns. Around every cistern along the route, there was a struggle for water until it became empty. The stronger of the men only got the water, and the weaker ones suffered much for the want of it. Many times, I saw men drink the muddy water from horses' foot tracks in the road. Some received sunstrokes for need of water and the intense heat.

No halt was made until midnight, and then we slept on arms.

July 9 – The column was in motion at 7 a.m. The heat was oppressive from the first. Provisions were short, and the anxiety for water was on the increase. There were more cases of sunstroke. A puddle of water was found with two hogs wallowing in it. The hogs were driven out and myself and others filled our canteens from the puddle. This may have been hard on the hogs, but I am suspicious that they did not need water to cool their blood long after discovery.

About sunset, our advance guard came upon the enemy's pickets about seven miles from Jackson. A halt for the night was ordered, and the men made themselves as comfortable as thirsty, hungry men could.

July 10 – The enemy's rear guard retired before our advance with little resistance. Our men were worked slowly and carefully forward, feeling the way up to hedgerows and patches of woodland to prevent surprise. At noon, a halt was made to allow the men to eat whatever of dinner they might be lucky enough to have. Not feeling the pursuit, the enemy advanced, but again retired at our approach with feeble resistance.

As Sherman's army approached Jackson, our brigade under Colonel Lesure of Pennsylvania, worked its way through thick undergrowth toward the north of the town to the insane asylum, a mile and a half from the city. The stars and stripes were run up on the asylum as the sun went down, and we slept on arms beside its walls.

Although our movements had not been conducted with great noise or much firing, the unusual bustle about the grounds set the lunatics wildly raving. As the windows of their cells were left open to give them needed air, we could see them and distinctly hear every word of their crazy talk. One gray-headed grandma talked the night through of John's wife and the children, while she knit upon the leg of a long blue stocking. A dark, full whiskered man gave the pedigree of all the horses and their drivers of great note and little note in the United States. A wife retailed all the particulars of a husband's infidelity and abuse, and so on, each had his illusive fancy. Their chatter was heartrending enough to drive a healthy brain to stark madness. It was not a pleasant night for us.

July 11, 1863 – The 2nd Michigan was detailed as skirmishers in front of its brigade and deployed as such near the Pearl River to the north of Jackson and facing that place. The orders were "Conform your movements to the 45th Pennsylvania on the right and go forward until you draw the fire of the enemy artillery." Before the sun had risen, the 2nd, deployed as skirmishers, were slowly sweeping in a long line the fields and woodlands, ravines, and ridges between the asylum and the city of Jackson. At first, a few scattering shots from the enemy's videttes was all the resistance me with. About six o'clock a.m., the line emerged from a wooded ravine onto an open space that rounded gracefully over into another wooded ravine, which was occupied by the enemy, and the slope that led up out of this ravine culminated in the high ground on which their fortifications were plainly visible. At this point, the advance was checked by a heavy musketry fire. For sometime, a lively fusillade was kept up across the ridge, but neither could tell if the other was being damaged.

In the meantime, the other regiments of the 2nd Brigade under Colonel Lesure had been massed together in the ravine in our rear. Feeling confident of good support, we were ready for anything that might come.

Then came the order "Forward double quick." At the command, every man sprang forward, discharging and loading his gun on the run, and every throat poured forth a yell that seemed to exceed the power of mortal man. A hailstorm of leaden hail from the enemy plowed the crest of the ridge and shrieked through the air as our heads appeared above the hill. Unchecked and undaunted, this human cyclone dashed straight upon the enemy's picket line, which broke and fled. The reserve picket fired one hasty volley and fled up the incline. Without a pause, this line of men five paces apart, straining every muscle of limb, plying the ramrod, and howling like mad creatures, assailed the enemy's line of battle and drove it inside the works. Then there was time to pause and look over the situation.

We were so close to the forts that our rifles prevented them from approaching their guns to fire them. Practically, their forts in our front were silenced, but our own situation was critical.

The 45th Pennsylvania on our right had gone but a few steps, and then fell back leaving a great gap on our right. There were no troops between us on the left and the Pearl River. There was nothing to prevent the enemy from marching around our either flank and drive us into their own camp. Our support did not come up and when Colonel Humphrey sent back for orders, it had been withdrawn, leaving us to face this flank of the rebel army alone. After an hour's delay, the 2nd fell back across the ridge to the former position. Twelve of our brave men had been killed, thirty-six wounded, and eight taken prisoner. And to what purpose?

In a congratulatory order, Colonel Humphrey says of this action: "This achievement you may well claim as among the most brilliant of the war. For a skirmish line, entirely without support, with no connection either on the right or left to charge an enemy drawn up in line of battle, drive him into his works, and charge those, is an unparalleled undertaking, an undertaking as audacious in conception as it was brilliant in execution."

An officer on the staff of General Sherman reports that he was with Sherman on the observatory of the lunatic asylum and saw the charge over the treetops. He says, "Sherman, when he saw it, swung his hat around his head three times and sailed it away."

A sharp firing was kept up across the ridge the remainder of the day and mostly all night.

An incident or two of the afternoon will show the spirit of the men.

Although the firing was general all along the line, there was one shot that we soon came to know from its different tone, and the fact that it seemed to plunge somewhat downward. The whole line was watching to see where it came from, but without avail. Finally, a marksman, Smith by name, was detailed to find and shoot the man with the privilege of going where he pleased to do so. Smith crept up and down the line briefly, exposing himself, until he had narrowly escaped being a victim several times. Believing he had located his man, he carefully aimed at a clump of thick foliage in the top of a tall elm tree and fired. With the crack of his gun, a figure came tumbling over and over out of the treetop, and we distinctly heard the "clung" as it struck the ground. That peculiar shot troubled us no more.

A venturesome soldier had crawled up to a large apple tree that stood near the crest of the ridge to get a better shot at "Johnny Reb" and had called three of his comrades to show what he could see. Of course, they were seen by the rebs, who opened fire. The first one stood up straight behind the body of the tree, and the next behind him, and so on, until all four were hugging as close behind the three as they could get. The firing concentrated on that tree, its limbs were clipped, and its body was filled with lead. Meantime, the quartets were convulsed with laughter at their situation and were powerless to act until someone called out, "The hind man falls over backward." This they did and crawled out of danger without injury, amid the merriment of the whole line.

The next morning, at daylight, we were relieved by another regiment, and we retired to a position near the asylum.

Rations and water were extremely scarce. Half rations of bread were issued and nothing else. Foragers were sent out to procure food, but returned with only a few ears of green corn. For the next five days, we were hungering and thirsting in camp, knowing but little of what was going on about Jackson, save that we heard the firing and occasionally one of the enemy's heavy shells would drop into our camp.

On July 17, the pickets discovered that the enemy's works were vacant and that Johnston had retreated across the Pearl River.

Many scenes of excess were enacted in Jackson before our forces had formally entered the town, and many acts of vandalism were perpetrated by the vast number of western soldiers that straggled into the city. In fact, the town was given over to sack and debauch for some hours.

CHAPTER 13 – JULY 17, 1863

With the news of the evacuation of Jackson, came orders to the 2nd to move at once. That night we camped ten miles out on the Canton Road, at Gray's Mills, near the Pearl River.

July 18 – Reaching Madison Station on the New Orleans, Memphis, and Great Northern Railroad early in the day, the work of destroying the railroad was begun. Several pieces of trestle work were fired, and the track on the solid ground was overturned by long lines of men, the spikes knocked out, the ties pulled up, and the iron rails placed on top. Then the ties were set on fire. When the heat became strong, the rails would squirm and twist like worms in the fire, rendering them useless until again passed through the rolling mill.

Rations being short at Jackson, there was issued but one day's rations at the outset, and none had been sent us and little was to be got in this country that had been acquired by both armies. Not only was the quartermaster anxiously searching for anything that the men could eat, but every man kept an eager lookout for signs of food. At every plantation, inquiry and search was made with no success. The men were working two hours on and two hours off. During any two hours off, I explored beyond the village to a very large plantation establishment at which were one surly white woman and about thirty young Negroes of both sexes. Inquiry for food brought the response that they had nothing for themselves. Thinking the white woman looked too healthy to be starving, I wandered away about the stables and Negro huts, until an intelligent contraband was found alone, who imparted the information that "De smoke house is plum full," and then vanished. Just by the edge of a grove in the rear of the house, I found chain and padlock. A heavy stock of cordwood, used as a battering ram, broke the chain at the first run, and as the door swung open, at least a ton of fine bacon and hams were exposed to view. Taking a nice ham as

130

evidence of what I had found, I started for the regiment to notify the commissary. Backward glances revealed many a wooly head whisking among the trees followed by hams and sides of bacon.

The commissary, ascertaining that the other smoke house was also full, took enough for our immediate needs.

There were no orders to destroy the station house but somehow if caught fire, as did a warehouse filled with cotton, and they burned to the ground, there being no conveniences for extinguishing the flames, nor any time to spare from the work we had in hand. The work of tearing up the track led us back toward Jackson and bivouac was made two miles out.

The next morning, the work was resumed and continued until noon when orders were received to return to Jackson, which place was reached after nightfall, distance thirteen miles. The principal part of our subsistence since July 12 had been green corn without salt. There was plenty of this, and no inconvenience was caused by it.

July 20 – Without rations, except the green corn and some melons and peaches found by the way, we began a march back to Milldale. The weather was intensely hot, and the unclouded sun glared down upon us with consuming rays.

The same struggles for water and the same suffering for want of it that met us on the way out, was gone through with again. Through all the privation, hunger, thirst, and terrible heat, there was no murmuring.

A halt was made at night eighteen miles from Jackson and four miles from Brownsville.

July 21 – March resumed. The ground was becoming parched up by the sun. The thermometer ranged up to 112 degrees in the shade. The men were being broiled alive and suffocated with dry hot dust.

Brownsville was passed with flying banners to show the people that it was no retreat. Men were choking with thirst and many were sun struck. An inquiry began to circulate gently among the men, "Why this haste on so hot a day?" More men dropped by the wayside. The demand to know "Why?" became more urgent, but none seemed to know. Then came the rumor that "Two generals had bet a basket of champagne that our corps would cross the Big Black in two days from Jackson." When this rumor had got well circulated, the troops had marched fifteen miles and were with one and a half miles of the Big Black at mid-afternoon. Just at this time. some obstruction at the front halted the column. The men sought shelter from the blazing sun under bushes, trees, and in the standing corn, anywhere for a little shade. In a moment, the mounted officers gave the customary order "Forward" and started their horses. Not a foot soldier responded to the order. Thinking the order had not been heard or understood, it was repeated with more force. And yet, there was no movement of troops. In blank astonishment, they contemplated for a moment the dry, dusty, hot, inactive silence, and then concluded that if the troops would not march with them, they had better bivouac with the troops. A victory for the men without a word or an act. To prove the truth of such an army rumor would be next to impossible, but subsequent events seem to confirm something of the sort. Any rate, the fact that the remainder of the march to Milldale was made by very easy stages, and then we remained at that place two weeks, inactive, proves that someone in authority was very foolish and cruel.

July 22 – Marched three miles, crossing the Big Black to the first place that would afford a camping place. After noon, there arose one of those sudden rainstorms peculiar to the locality that cooled and cleansed the atmosphere.

The next morning was bright and balmy and while the earth was still damp and the air refreshing, we arrived at our old camp at Milldale. The little level plateau in the bend of the old creek bed received the 2nd Michigan again as its tenant. The recent rains had washed away the soapsuds, leaving a limited supply of rainwater for washing purposes. The old tramp of two miles to the spring in the dell for water to drink was resumed, rations were plenty, needed clothing was issued, no arduous duty was performed, and it was rumored that our next move would be up the Mississippi. For two weeks this lazy life was led, enlivened only by such incidents as are common to camp life and a heavy draft on the regiment.

July 28 – A call for men to guard prisoners to Indianapolis. As all were anxious to get out of this desolate country, this was no hardship. A pugilistic drill between two members of the same company, each accusing the other of impudence, enlivened one day. As all seemed to think they were both right, they were allowed to settle the matter between them without referees or assistance from anyone.

One day as I was seated upon the grass, enjoying some sweet potatoes and peaches that I had brought from the farm—I don't know whose—with some wheat bread, furnished by the government, a young contraband came into camp and wanted to "Hire a boss." Some wag referred him to me. His account of his accomplishments were satisfactory to me and the question of wags, a place to sleep and other matters were agreeable to him, until I asked him if he could live on what I was then eating. With a look of lofty disdain and highly elevated nose, he said, "I must have cawn bread, bacon, and Irish 'taotes, or I can't work for you." These hard conditions, I have no doubt, lost me a valuable servant. The easy idle life we were leading lasted until orders came to be ready to move at a minute's notice—"leave not to return."

August 3 – Late in the afternoon, we broke camp and marched to Snyder's Bluff. The steamer lying at the landing seemed to assure us that we were bound north. The commodious steamboat Ohio Belle, bound for Cairo, Illinois, received the 2nd Michigan but did not move from the landing during the night.

August 4 – With the early daylight, everyone was on the alert. Moorings were cast off, the ponderous engines began to move, and the Ohio Belle soon left the slimy, green, sluggish Yazzo, and turned her prow up the broad swift current of the Mississippi. The sun was bright, the breezes from off the rippling waters were cool and refreshing, and our hearts were buoyant for we were leaving behind the memory of the toil, hunger, thirst, and suffocation of the last two months. What our future destination might be was of little concern. This move was a temporary relief, at least, and each seemed to be intent on its enjoyment.

The passage up the river was characterized by no important event although a movement of live troops will always be full of incidents.

Stopping to coal at Helena, Arkansas, on August 5, many of the men went ashore. When they came aboard, each had a quantity of necessities or luxuries, which they had "purchased" on shore. The list comprised of underclothing, cigars, eatables, shoes, tobacco, and other things. They must have had great credit in this southern town for money was very scarce with all of us at this time.

This being the dry season, the river was shallow, and it was necessary to have a man at the bow taking soundings a considerable part of the time, especially at night. The peculiar drawling chant of the leadsman and his quaint language was a source of much amusement to the soldiers. As they lay rolled in their blankets, apparently fast asleep, and the stillness of night was over all nature, the leadsman, with lead and line, would cast the lead far ahead and note the number of knots the depth of water showed, he would chant the measurements to the pilot—in half three three, quarter less three, 'n half twain, by the mark twain, ending each call with a sudden jerk. From some part of the slumbering host would come "knee deep" in exact imitation of the leadsman. When the culprit was sought, the cry would come from the other end of the deck from the search, and so on with no results for everyone was asleep. This was very annoying to the boatman, but fun for the boys. At two o'clock a.m., August 6, the boat grounded on a sandbar about fifty miles below Memphis, Tennessee, and sprang a leak. After a time, the boat was worked off the bar and reached Memphis at 4 a.m. To do this, the pumps had to be kept working.

At Memphis, the troops were unloaded while the boat was repaired. Arms were stacked upon the levee, ranks were broken, and no guard was posted. Turned loose, as it were, in this large city after a campaign of privation and want, the soldiers had full possession of the city before morning, and the civil authorities, nor the provost guard, could help themselves. The soldiers would go by squads into restaurants and order sumptuous meals, eat them leisurely, request the proprietor to charge it and depart. Fruit stands and bakeries contributed to the general entertainment.

Brilliantly attired military officers, at the head of the colored soldiers, trying to make arrests, were dumped into the gutter like rubbish, and their guards chased out of the city. To the credit of the men, there were no indignities offered to civilians, although this was known to be one of the bitterest secession towns in the country.

August 7 – Embarked again on the Ohio Belle at sunset. Many of the men had to be brought in under guard, but we left Memphis with all on board.

The next morning, while steaming up the river at a satisfactory rate, I received orders to report to the colonel in the cabin. Surprised by the order and wondering what part of the irregularities of yesterday I was responsible for, and if the punishment would be severe, I reported at once. Entering the cabin and saluting the Colonel, he said to me, "Do you see those candy jars?" pointing to a row on the side of the cabin.

"Yes, Sir."

"Well, they are full of brandy. Some of the men smuggled them aboard yesterday, and we captured them. Take them on deck. Let each man on the boat have one gill and no more, to be drank in your presence."

The Commissary, whose duty it was to issue all rations, liquor as well as others, was present, but there was nothing for me but to obey. The eight gallons of prime brandy was disposed of agreeable to the colonel's orders, and no one got enough to produce drunkenness.

August 9 – We arrived at Cairo and took the Illinois Central Railroad.

August 10 – Breakfasted at Centralia, changed cars to the Ohio and Mississippi Road at Sandoval, supper was had at Vincennes, breakfast at Seymour and arrived at Cincinnati, Ohio at noon. Dinner was served to us in the market shed, and we crossed to Covington, Kentucky, and took up quarters in the sheds of a rolling mill near the ferry dock. The rolling mill was in full operation, and I witnessed for the first time the process of manufacture of T rails. While some were smelting the ore and running it into molds called "pigs," others were heating the pigs and running them through rollers to make a rail. Each heating of the iron was allowed a day to cool, but all of the processes were going on at once in different parts of the mill. As the iron went through each process, it was corded up like wood to cool gradually. By the last set of rollers, the rail was given its T shape, and it was sawed off while red hot to the proper length.

The open mouths of the blazing furnace, the lurid glare of the hot iron, and the sweating bodies of the men, naked waisted as they handled the heavy iron with great tongs, slings, and other apparatus, formed a weird scene long to be remembered by one unused to such things. For men, who had for months been used only to the hardships of camp and field, it was a great thing to be in the heart of a great city.

In strolling about the streets, chance led me opposite an ice cream saloon, brilliantly lighted, with six handsome young lady waitresses, all in white. While gazing across the street at this enchanting scene with a great longing to join in festive society, a comrade joined me. For a time, we gazed on the enchanting scene, and finally resolved to pool our money and either buy out the concern or freeze up the waiters. To decide was to act. Entering the saloon, we prepared to buy some cream if the ladies would join us in eating it. Our offer was accepted, and at the end of two hours eating and chatting, the young ladies in white dresses were veritable icicles. Their teeth chattered as they asked us to call again, and we betook ourselves to a pile of hot bars of iron to get warm.

It was the coolest August evening of that year in Covington.

In the morning, we visited the glassworks and were introduced to the mysteries of blowing lamp chimneys, bottles, lamp globes, and other glass household articles.

Then the pottery works took our attention for a few moments. There was not much of interest in the shaping and baking of bowls and pitchers, but a woman and her five children who were preparing the clay attracted our attention and aroused our sympathy. These were ranged along a long table, from the mother to a little mite of a boy. The little fellow took some of the dry clay, poured water upon it and worked at it with his fingers until it was thoroughly wet, then the next one took it and mixed it a little more, each in turn working the clay into a finer paste, finishing with the mother, who worked it thoroughly with her fingers until it was properly prepared for the molders. This little brood, perched upon high stools, mixing clay with their fingers for their daily bread, touched a sympathetic chord in the breast of every soldier who saw them.

August 12 – Late in the afternoon, we left Covington by the Kentucky Central Railroad and breakfasted the next morning at Lexington. There was nothing about the town to recommend it except Henry Clay's monument.

At noon, Nicholasville, the end of the railroad, was reached. The place is a mere hamlet, though the surrounding country is fine and future years may develop a prosperous town. Bivouac was made in a little grove, outside the town, for the night.

August 14 – Camp was fixed about three miles out of Nicholasville and was called Camp Park. The men who were sent up the river from Milldale to guard prisoners rejoined us here.

After the camp was located, and we had reached it, the remainder of the day was spent in fixing camp, looking up the needs of the men in the way of clothing, provisions, and ammunition.

August 15 – A little incident occurred that well shows some of the arrogance of the superiors in the army and a little of the stubbornness of subordinates. Notwithstanding, we had moved by steamboat, rail, or on foot every day for eleven days, the Adjutant ordered a guard mount with clean muskets, white gloves, and shoes blacked. The shoe blacking was of little trouble. The white gloves were forth coming although not very clean, but there was not a clean musket in my company and not a minute to clean them. I marched my detail to the guard mount with muskets as they were and aligned them with the details from the other companies. Upon inspection, the adjutant promptly rejected them and ordered me to march them off parade. I led them to their quarters and directed them to take their guns under a neighboring tree, take them all apart, and renovate them in a first-class manner. Meantime, I was attending to my other duties, and the guard mount was waiting. Presently, the sergeant major came and informed me that the adjutant was waiting for me. I told him I had no other men than those I had offered, and they were cleaning their guns. He reported to the adjutant and immediately returned with a request from the adjutant that I bring them on, and they would be accepted. With great deliberation, the men put together their guns, and we got through that guard mount an hour late.

Though this was a little triumph, yet I was very careful that every musket was cleaned that day.

That evening, the camp being in good shape, my captain said to me, "You are as near your home as you probably will be during your term of service, and our major is the highest officer in camp of the 9th Army Corps. You can get a furlough in a few hours that will take weeks at any other time." Accordingly, I immediately made application for fourteen days leave of absence, the greatest length of time allowable by the rules.

August 16 – Just as the daylight began to fade, my furlough was returned approved. As every hour's delay took an hour from my leave of absence, I started at once for Nicholasville, where I was fortunate enough to catch an empty train starting for Covington. The next morning, clean clothing was procured in Cincinnati. At eleven o'clock, we were speeding northward toward Indianapolis, I having engaged to make a slight detour to carry $350 to the family of my captain at Galesburg, Michigan. Several comrades made me their money messenger. Altogether, I had on my person a thousand dollars of other men's money to be disposed of at various points. A three-hour wait between trains gave me an opportunity to look over Indianapolis that afternoon, and the next morning, I took breakfast at Michigan City, Indiana. At that place, a whole trainload of Confederate prisoners passed while I waited for my train. They all had a word, good, bad, and indifferent for the lone Yank standing upon the platform.

Later in the day, when I left the train at Galesburg and inquired for Captain Stevenson's, the woman and children followed to learn who it was and what news I brought. A short two hours passed in which I was required to eat more than was necessary and to tell all I could about the boys in camp. When the next train arrived, these people were loath to let me go.

They had not heard all they wished to hear. Still later in the day, as I stepped from the train in Jackson, Michigan, I stood before my father. I had to tell him over and over who I was before he could realize that it was really I.

As there was no train for home, we stopped at the Hibbard House over night. During the evening, several old schoolmates who were in the city, came to see me. The extreme meanness of the "stay-at-homes" was exhibited by one of these. Seeking me alone, he expressed great surprise that I should be approaching my home by a round about way, and said if my papers were not all right there was great danger, and he would do anything to help me. Being assured that my papers were in proper shape and that I would exhibit them when anyone having proper authority to demand should require it, he soon left me. My suspicions were aroused and a little inquiry revealed the fact that he made it his business to pick up soldiers whose papers were not all right, for the petty reward the government paid for each such dirty work. The remainder of August was spent in visiting friends and relatives.

September 2 – Leave of absence being expired, I started on my return to my regiment, which had moved to Crab Orchard, Kentucky. A lady and daughter seated opposite me on the Cincinnati, Hamilton, and Dayton train illustrated the strong interest the loyal ladies felt in the soldiers. My complexion being fair, and having been up all night before, I looked somewhat pale and careworn. This attracted the lady's attention, and she urged me to share their lunch. There being some nice cake and sandwiches left, these I was compelled to take or seem rude, as she assured me that she and daughter would be at home to tea, and that I hardly looked able to endure the hardships of the army. Finally, pressing upon my acceptance the napkins that covered their lunch, they left the train with many regrets that I should have to return to the army before I was able. As all of this was inferred without word of, I could not bear to undeceive the, and thus at some other time rob a needy soldier of assistance.

CHAPTER 14 – SEPTEMBER 4, 1863

This day found the 9th Corps encamped near Crab Orchard, Kentucky.

This was once a famous watering place and health resort. Mineral springs of some excellence are here, and the hotels that once accommodated visitors to the springs still stand though somewhat dilapidated, as evidence of the fashionable gatherings of health and pleasure seekers, who whiled away their summer leisure amid the verdant hills and woody dells of the neighborhood.

The surrounding country is picturesquely rolling, swelling gradually into grander hills to the eastward, until far toward the rising sun, the blue line of the Cumberland Mountains is plainly visible. The beautiful scenery, the pure mountain air, the genial sun, and the curative properties of the spring water, make it a desirable resort for invalids.

But what has a soldier, just returned from a leave of absence, to do with these things? Work is now before us. One used to camp life would predict an important move on foot by the work being done by the troops. The men were supplied with all needed clothing and equipments of every kind. Thorough inspections of the troops were made. Broken and worn arms and trappings were condemned and turned over to the Quartermaster. All extra baggage was ordered packed for storage, and finally everything that could be construed to be superfluous must be disposed of in some way. When it is known that the orderly sergeant must ascertain just what each man of his company needs, from a pair of socks to an overcoat, and from a cone pick to a musket, consolidate all the needs and make a formal requisition for just the number of each article required, and then see that each man gets just what he requires, while no requirement of the daily routine of camp must be neglected, it will be seen that there is work for him to do at such times. Six days of this turmoil and confusion sandwiched in between reveille, roll call, sick call, morning report, guard mount, police call, drill call, dinner, afternoon drill, dress parade, and tattoo roll call, immediately followed my leave of absence.

September 9 – All the orderly sergeants heaved a sigh of relief when the order came, "Be ready to march tomorrow with four days rations, without extra baggage, leave not to return."

Our destination was supposed to be East Tennessee, but the enlisted soldier knows nothing of his future but such as he can guess.

September 10 – Reveille sounded at four o'clock. As soon as breakfast could be got out of the way and camp broken up, we were on the move southward on the macadamized turnpike. The country was uneven and somewhat broken. In many places, there were long stretches of newly broken stone on the turnpike, and the sharp corners were bad work with the feet and shoes of the troop. An easy march of eleven miles brought us in sight of Mt. Vernon, Kentucky.

September 11 – Passing directly through Mt. Vernon, we were afforded a cursory view of a southern county seat. The place contained perhaps thirty dwellings, counting log houses, huts, hovels, shanties, and pigsties. A limited number of houses were frame buildings, devoid of paint or whitewash, sadly deficient in clapboards and wanting many a shingle. Those houses that could boast of anything neat and tidy were flanked on either hand by low dirty tenements, occupied by Negroes and poor whites. Surrounded as the place is by a fertile country, there appears on the surface no other reason for the condition of the village than the blight of slavery. Dependence upon a degraded and despised race for everything pertaining to their subsistence seems to destroy all the energy of the whites and cause them to choose squalidness without labor, rather than neatness and thrift with labor.

To the southward of Mt. Vernon, the road led through an uneven and broken country. At some points, rocks rose on either hand to a considerable height, while at other places, two hundred feet deep on the other. The entrance to several caves was passed, and the noon halt was made on the banks of the Wildcat Creek. Fording the stream, we traversed the rugged road over Wildcat Mountain. Though we were traveling on a roll road, there were places where the wagons had to be eased down the steep descents by a long line to the rear with forty or fifty soldiers holding back. At one place, there was a perpendicular drop of three feet in the solid rock in the wagon track. At another place, two wagons, with mules and freight, were precipitated down the steep declivity of two hundred feet at the side of the road and were a total loss. The men were heavily loaded with their own necessary supplies, water was scarce, and they were compelled to help the mules pull their loads up the hills and hold back for them downhill. Many of the weaker ones tired out. Fifteen miles of this toil brought us to a little level bottom of Rockcastle Creek, the only level place we had seen during the day big enough for a camp. Here was plenty of good sweet water, a great blessing to the weary and thirsty.

September 12 – At daybreak, the column was in motion. The first mile was a rapid ascent to the top of a narrow ridge called the Hogback. The road lay on the top of this hogback for nine miles. In many places, there was barely room for the wagon track, while on either hand there were wide and deep ravines. A number of baggage wagons were upset and some were precipitated down the steep declivities, destroying the wagons and freight, and killing the mules.

At eleven o'clock, we halted, it being Saturday, to remain until Monday.

The few natives who visited us were of the long, lank, lantern-jawed species, with clay colored complexion and tawny locks. They called themselves white and were too proud to work—"work was for niggers"— yet they reveled in all the pride of poverty of the most abject. Shiftless, licentious, malicious, and unclean, as a rule, could be applied to both sexes. The sight of money aroused a desire to possess, either with or without an equivalent. The limited opportunity offered by a Sabbath among these people seemed to demonstrate that honesty, honor, or virtue was unknown among them. Loyalty to the Confederacy and the state of Kentucky appeared to be the sum of their sentimentality.

Some of the men who pitched their tents on low ground were driven to other shelter by drenching rains. The Sabbath was utilized by the paymaster to pay as many as possible.

September 14 – Early in the day, the line of march led through London, the county town of Laurel County. The road upon which we moved being the only street of the town, we saw all there was of it. Some of the buildings were of brick and passably neat, but the frame never knew paint.

The country south of London is mostly timbered, interspersed with small plantations under cultivation, and is less broken than that before passed. The inhabitants were more intelligent and thrifty and less vicious. Having moved fifteen miles, camp was pitched at three o'clock.

September 15 – At five o'clock, the morning meal being over, the march was resumed. The country was variegated, but the march was easy. In the center of a little valley, we came upon Barboursville, the county town of Knox County, which closely resembles London. After a march of fourteen miles, we camped upon the banks of the Cumberland River, a mile beyond Barboursville.

September 16 – Following the Cumberland eight miles, camp was made upon a triangular plain enclosed by three ranges of hills. On the north, a precipitous range of hills, rising to a height of three hundred feet, inclining southeasterly seems to meet a similar range, inclining northeasterly on the south. West of the Cumberland, another range seems to connect the two, thus forming a complete triangular enclosure. That the enclosure is not complete is proven by the fact that the river enters at the northeast angle.

The rations that we started with were now exhausted. Heavy rains began to fall, and we are in a hungry rainy camp to wait until rations can come up from the rear over rough and rugged roads we have left behind. In this sparsely settled section, it is a miracle for an individual to obtain a meal of victuals from the inhabitants. They never have a surplus of provisions or any thing else except it be children. An incident will illustrate the character and habits of these people.

During our two days' stay at this camp, taking a comrade with me, we set out to find where a footpath led to that ran through the undergrowth not far from our camp. Following the trail on and on, it led us higher and higher up the mountain. After a brisk walk of three or four miles, we at last came to an opening of about three acres with a cabin in the center. One-half the opening was planted to corn and the other half to tobacco. At the cabin, we found a woman and twelve children, the children ranging from one to twelve years of age. We were cordially welcomed, and the conversation elicited the following facts. This family lived off the product of this three acres of corn and tobacco, supplemented by what the father could hunt and fish for. The corn was grated, while still on the cob, into a coarse syrup by the older children, and, said the woman, it keeps some of them grating all the time. This they ate. The tobacco, when cured, was "toted" to the nearest market and sold to buy the absolutely necessary eatables and supply their scanty wardrobe. And scant it was. The woman's only garment was a dirty calico dress reaching halfway between the knees and ankles. The feet had never seen shoes, and present indications suggested that they were unacquainted with water. A similar loose gown covered some of the children while others had less clothing on than a government mule has when he is in harness.

When we had learned about all we could of their manner of life, my loquacious comrade entered into a glowing description of our northern grain fields, harvesting machinery, grist mills, and labor-saving machines in general. The woman drank it all in with open eyes and open-mouthed wonder until he described a dog churn. She then burst into a derisive laugh and said she now knew he had been lying all the time, for they could not fix it so a dog could churn.

Though there was a four-story gristmill twelve miles away over the mountain on the main road, this woman never had seen one. In ignorance, surrounded by squalor and dirt and in want of any comforts of life, this mother of twelve, clay-colored, tow-headed dirty urchins, thought her house the most glorious the sun shone on.

September 19 – The column was again put in motion. Fording the Cumberland and following its banks a short distance, we passed between two quite lofty peaks between which it passes. Rising steep and sharp on one hand, a sugar loaf leans toward its fellow. On the other hand, the mountain rises as abruptly, but it overhangs like a man with his head drooping forward. These two peaks seem to have the attitude of acknowledging an introduction. On a little tributary of the river stands a gristmill characteristic of this section. Four saplings are planted in a square form, cross poles are laid on, and on these is a bark roof. Beneath this is a single run of stone boxed in with an outlet on one side only. A funnel-shaped hopper to run the grain into the stone and a wooden spout to carry the meal from the stone into a cooling trough. The turbine wheel directly under the stone furnishes the power, and that is all there is of it.

Ten miles of indifferent country, and then we camped near the base of some picturesque craggy mountains that were on the west from the road.

Provisions being scarce, some hunger was experienced. Perhaps it was this that caused us to camp before noon.

I felt an overpowering desire to get a view from the mountains. Their sides were steep, and in places, large squares of gray rock with perpendicular faces stood out in bold relief from the stunted undergrowth that covered the mountainside where there was sufficient earth to sustain it. The climb was soon begun with two comrades. A height of perhaps four hundred feet was attained by a persistent effort. Most of the way, we pulled ourselves up by aid of the roots and limbs of the undergrowth. At this height, we could look over lesser hills and see the main charm of the Cumberland Mountains, and catch glimpses of the main road as it wound its way over hills and down dales, through the ever-varying landscape. The view was magnificent, but a little valley down the other side of the ridge on which we had rested attracted our attention. Completely land-locked by rough hills was a valley about large enough for three farms, and at some distance from each other, were three log cabins. Love of adventure and a stray thought of dinner decided us to descend and pay a visit to three sequestered people. We reached the nearest log house about noon and made a request for dinner. If we could wait for it to be cooked, we could be accommodated. Would we like chicken? Certainly. Then we stretched out on the grass where we could see through the open door of the cabin everything that transpired.

In the open fireplace, water was put to boil in a large skillet, while a chicken was run down and killed. The chicken was scalded in the skillet. While the chicken was being picked and cleaned, potatoes were put to boil in the skillet. When the potatoes were done, they were removed from the skillet and the chicken took their place. While the chicken was roasting, the potatoes were mashed in a tin pan and put by the fire to keep warm. Biscuits were mixed ready to go into the skillet when the chicken came out.

The biscuits were kept warm by the fire while tea water was boiled in the skillet. This ended the cooking and dinner was served at three-thirty o'clock with tomatoes, onions, and buttermilk as relishes. During the preparation of this meal, the presiding genius of the place, a black-eyed, black-haired mountain maid, Maggie Sampson, had played a tattoo with her coarse shoes on the floor of the cabin as she tripped to and fro about the work in hand.

Hemmed in by mountains, these honest hardy mountaineers know little of the world of strife, lived here content and happy, not knowing or caring about the ways of the outside world. Our meal finished, we left them a half dollar each—which they were loath to take—and began our climb toward camp. They told us it was four miles by the way we came. By the wagon road, it would be twenty. Before darkness set in, we were safe in camp and ready for a good sound sleep. It had been better for young athletes as we were to take this invigorating climb than to set brooding over the delay of the trains with our rations. Besides, we had a good dinner.

September 20 – The supply trains, having caught up with us, the march was resumed in the early part of the day. The hilly roads were soon left behind, and we emerged upon a more level country as we approached the base of the Cumberland Mountains. The last five miles before reaching the base of the mountains, is comparatively level, and some of the land is under cultivation though much of the country is of poor soil.

The road leads diagonally up the mountainside by a gradual ascent until the center of Cumberland Gap is reached. At the summit of the pass, the jagged and overhanging rocks rise on either hand several hundred feet. A large square stone planted at the corner of Kentucky and Virginia and on the line of Tennessee located near the center of the gap. Without a halt, we moved on beginning the descent, and after a march of twelve miles, we went into camp on the mountainside two miles beyond Cumberland Gap in the state of Tennessee. The grandly undulating hills, dwindling to lesser undulations in the distance, with the ever-varying changes of light and shadow, the gray rocks, the green fields, and the paw paw groves tinged with gold, formed a beautiful picture of the broad sweep of vision, commanded by our elevated position. The paw paws themselves were ripe and were larger, fatter, and yellower than those grown in a colder climate. Many a soldier filled the void left by our scant rations with the fruit of the paw paw, for, be it remembered, we were on short rations and were likely to be so indefinitely.

A three-story gristmill in the distance was evidence of a higher civilization than that left on the other side of the mountain.

September 21 – The onward march led us through some fertile valleys and over some high hills in the gradual descent. Powell's River was crossed by means of a primitive ferry. A large rope tied to trees on opposite banks of the river and a flat scow was the ferry. When the scow was loaded, you reached the opposite shore by pulling on the rope. The little boat plied rapidly back and forth across this clear and swift stream, carrying fifty men or one wagon at a load. In this way, the troops crossed much quicker than one would think. The noon halt was made at Tazewell, the county town of Claybourne County. This was once a thriving village situated on the border of a fertile valley in the midst of a community of honest, industrious farmers, who, though somewhat ignorant of the outside world, were polite and hospitable in their rude way. The town was in ruins from the torch of the rebel army in its last retreat from Cumberland Gap. Fearing that it might shelter the Yankees, the rebels left all the principal buildings a heap of blackened ruins. As we passed through the place, a young lady with a fine intelligent face said, "I wish you-ones would set the drums beating for I never heard one in all my born days."

At Tazewell, we left the Knoxville Road and took that to Morristown. This road led through a very rolling but cultivated country. This day's march marked eighteen days of progress.

September 22 – Early in the day, Clinch River was forded, and we approached the foot of Clinch Mountain. The road leads over the mountain through Thornhill Gap and rough and rugged road it is.

The half-fed soldiers, with all their personal baggage on their backs and their muskets also, were strung along beside the mules as close as they could march, pulling with the mules by the traces to drag the heavy wagons and artillery up the rocky way. Six weary hours were consumed in this way on two miles of road. From the summit of Thornhill Gap, a wide view of a wild and rocky country was had. The thought intruded, *is there no rest from hills and rocks? When will this tugging cease? And when will we get a square meal?* The descent was not so toilsome, yet the wagons had to be eased down the steep way by long ropes in the hands of soldiers. Fourteen miles of this toilsome march, and we rested on the banks of the Holston River.

The broad valley of Holston, dotted over with its plantations as it stretched away to the west between mountain ranges, was pleasant to look upon after the wild scenery left behind.

September 23 – The Holston was forded, being the fifth river in our course without a bridge, and a march of six miles brought us to Morristown, county seat of Hamblin County on the line of the East Tennessee and Virginia Railroad.

The town was a very unpretending place of not more than a hundred houses. Here we rested for the rest of the day. The first thought in these days of hard work and light diet was to get something to eat. Soon as camp was established, I started out with a comrade to see if we could get supper among the inhabitants. As every man who had any money was eager to procure food, a little strategy was necessary to secure a meal in comfort. Avoiding all places where soldiers were about, we passed to the edge of the town opposite the camp. Seeing two horses with side saddles hitched to a gatepost, we decided to call at that house. Approaching the open door through which we could see four ladies seated in the parlor, we rapped for admission. Reluctantly, the lady of the house came forward. To our request for permission to come in, she replied, "You will come in if you want to, with or without permission. Assuring her we were only in quest of supper and would retire if it was not agreeable to her, she thought a little and asked us to take seats. Then the visiting ladies commenced a bitter attack upon the "Yankees." There were no gentlemen in the Yankee army. They were roughs, villains, cutthroats, and the off scouring of the earth. Keeping our tempers placid and laughing at their bitter expressions until their vocabulary was exhausted, we finally got our chance.

"Was she acquainted with any Yankees?"

"No!" She had no acquaintances but gentlemen.

"Were all the Confederates gentlemen?"

"Yes, every one."

"Think again. Is there not at least one man in the Confederate army whom you would hesitate to associate with?"

"Well, yes, perhaps one."

"Now, really isn't there many?"

"Well, I'll be honest with you. There are many, but most of them are gentlemen."

"That is exactly the case with the Yankee army. The great majority of its numbers are gentlemen, but it is to be regretted that a few are not, and if you ever become acquainted you will find this statement true."

After this conversation ran more smoothly. At last, the visitors arose to depart, and we assisted them to their horses. Before they rode away, they said they were glad to have met two Yankee gentlemen and invited us to call on them should our stay permit. During supper, jokes were passed freely upon both the Union and Confederate armies. After supper, the sister of the hostess was asked to play upon the piano. She would be glad to, but she knew none but Confederate songs. Being informed that these would not give offense, the played the *Confederate Wagon,* the *Bonnie Blue Flag* and others. After we had listened to these, she whirled upon the piano stool and hesitatingly said, "You have been so kind, I think I will play the *Star Spangled Banner* for you." Which she did and was rewarded by three rousing cheers from soldiers who had congregated in the yard under cover of the gathering darkness. I afterwards learned that the woman's husband was a Confederate officer, then a prisoner of war in Fort Lafayette.

September 24 – It was the intention to transport the troops from this place by rail to Knoxville, forty-two miles to the west, but for want of cars, our brigade was compelled to march this distance. The wagon road and railroad both run westward near the center of the valley of the Holston, which broadens to nearly forty miles in width. Holston Valley is the garden of the state of Tennessee. Its inhabitants are industrious, homespun people, mostly loyal to the Union, with an intensely bitter hatred of rebels. With a healthy climate, active habits, and plain wholesome diet, they are a robust, hearty people.

We rested at night at Mossy Creek, a small railroad station.

September 25 – At Strawberry Plains, the road recrosses the Holston. Here a blockhouse had been erected to defend the railroad bridge. At the blockhouse, the people were assembling to hold what they called a "barbecue" but what was in reality a basket picnic, in celebration of the reoccupation of East Tennessee by Union forces. The variety and antiquity of styles of dress, their hearty awkward manners, their joyous welcome of Union troops and their bitter hatred of secession all entered into the composition of this grotesque gathering and made it worthy of study. But the soldier cannot choose his time for observation. Crossing the river by the railroad bridge, which the rebels had burnt, but the Union forces had rebuilt, we left these people to enjoy their feast by themselves, and we moved on toward Knoxville.

This portion of Tennessee had alternately been in possession of the rebel and Union forces several times. Each return of the rebels had been characterized by cruelty, murder, and a heavy draft upon the resources of the country. They had been a terrible scourge to these intensely loyal people, and now they hoped the Union forces were large enough to forever free them from their dearest foes.

September 26 – At noon, we arrived at Knoxville, camped upon rolling ground on the riverbank near the city. The rugged nature of the roads and the arduous work required on this march of 175 miles, together with the short allowance of food left us footsore, ragged, and emaciated of body. We were far from our base of supplies, all supplies of every description must come to us by wagon and pack mule over the road we had traversed. Half rations only were issued. Work was expected of us, and there was no shirking or complaining.

CHAPTER 15 – SEPTEMBER 27, 1863

Knoxville stands upon the north bank of the Holston River, about two miles from the eastern end of a natural ridge that extends along the Holston for twenty-five miles. This ridge is generally about one hundred and fifty feet above the river, but at many points, there are hills rising much higher. At Knoxville, this ridge is about a mile wide. Back of this ridge is a valley through which runs the East Tennessee, Virginia, and Georgia Railroad. At intervals, small creeks cut through the ridge from the valley to the river. Three of these run through Knoxville, and beginning at the east, are called First, Second, and Third creeks. The main portion of the city is between First and Second creeks, the valley, and the river.

Traversing the city from east to west, there is much ascending and descending of hills. The high points in Knoxville are Temperance Hill in East Knoxville, 225 feet above the river, Mabry's Hill, 230 feet, College Hill west of Knoxville, 160 feet, and a ridge between Second and Third Creeks with a general elevation of 200 feet. South of the Holston are a succession of elevations, the highest of which is 360 feet above the river. A few miles below Knoxville, the Holston and Little Tennessee rivers unite and form the Tennessee River. This place is situated upon a direct line of railroad connecting the eastern and western rebel armies, with all of these natural defensive positions, is to be the bone of contention.

Open to attack from east, west, and south, Burnside, with his 13,000 men surrounded by mountain ranges and a long way from his base of supplies, is expected to hold this section and prevent a repetition of the outrages and butcheries that have been heaped upon these intensely loyal people. The streets of Knoxville have run with the blood of Unionists riddled with bullets for daring to assert their love for this country. Citizens have been besieged in their own homes and shot at sight on suspicion. Others were hung without trial. Families were burned out of their homes and secession hatred caused many to languish in the loathsome jail because they would not espouse the rebel cause.

Inspired by these bloody outrages, the Unionists of East Tennessee were bitter, implacable, and bloodthirsty. The material aid and timely warnings of the inhabitants played an important part in holding this position against the superior numbers that could be brought against it.

Burnsides' army was constantly alert. Orders were issued to be ready to march at any minute. Rumors reached us that the enemy had crossed the Little Tennessee River and were advancing.

September 28 – Our brigade was ferried across to the south side of the Holston and took position on high ground two miles out on Marysville Road, to check any advance from that direction.

September 29 – Refugees were coming in all day, bringing conflicting and unreliable reports of the numbers and position of the enemy. Many of them were very much frightened. It was a pitiable sight to see these men, women, and children fleeing from good homes, with only such things as could be packed upon an old rackabones of a horse, going they knew not where, or when they would return. It was no uncommon sight to see a woman or a small child upon an old horse packed around with quilts, feather beds, cooking utensils, hams, and sides of bacon. The old and feeble, the young and helpless, and delicate women deserting the homes of civilization for the life of primitive savages. There was no shelter for them. They must take their chances beneath the skies broad arch with such protection as their limited personal effects afforded them. Captain Bryson, a fine specimen of the grisly mountaineer, came in with a hundred volunteers for our army. They had crossed the mountains from North Carolina, and made their way to our camp. A fine lot of hardy mountaineers they were, sharp eyes, and ready hand. No enemy appeared, and we pitched our poncho and made us as comfortable as half-fed men could be in such a position.

September 30 – The refugees still kept coming, but no enemy appeared. During the day, gathering clouds indicated an approaching storm. At four o'clock, the 2nd Michigan went on picket some distance to the front, and about the same time, heavy rain began to fall. Throughout the long night, the chilly, drenching rain continued to come down on the unprotected pickets. Anxiously watchful of the unknown country in front, every sway of a bush is scrutinized through the deep gloom, every sound is caught by the listening ear. An enemy might approach under cover of the black darkness, and the roar of the falling rain, but he will not surprise this picket. Realizing the importance of this duty, the sharp eye is always open, the quick ear is always strained. Yet he is not afraid. Being relieved of his two hours' watch, he retires a few feet to the rear, draws his set blanket a little closer, leans against a tree, and forgetful of hardship and dangers is lost in thought to of the last strains of music he heard at home, *"Heavily falls the rain; Wild are the breezes tonight."* [**From a Civil War song, *Brave Boys Are They*]** and the longing of a half-appeased appetite brings visions of the well-filled tables of friends at home. In sleepless dreams, the night wears away.

October 1 – More rain, more refugees, more tales of outrages, more rumors of advancing enemies, and more hunger. The inclement weather, scarcity of provisions, thin and ragged clothing, and the refugees all helped to make our position unpleasant. In other words, we had an extremely important duty to perform with almost everything against us. No matter how much the cold rain and our half-empty stomachs annoyed us, the front must be carefully watched, and every refugee must be carefully questioned and forwarded to headquarters. Just before dark, we were relieved from the outpost and returned to camp. For a week, there was comparative quiet. The stream of refugees gradually grew less and finally ceased. No enemy had shown himself, the people became less frightened and began to return to their homes. Many old men and boys offered to assist in the defense of East Tennessee. This led to the forming of the National Guard of East Tennessee on the following plan. The men were organized by neighborhoods into companies under command of some reliable one of their number as captain. They came inside the lines, were sworn into the U.S. service, each was given a gun and ammunition, and each returned to his own house. Their plan of operation was to be as follows: the first to see the advancing enemy would mount his horse, give the alarm to as many of his company as he could on his way directly to the Union lines. The others would rally as many men as possible at the first narrow pass in the road and bushwhack the enemy's advance. The effect of this would be to give us time to meet them fully prepared to give them a warm reception. Their aid in this direction and their assistance in procuring food for the troops was considerable.

October 8 – We had been notified overnight that we would march at 7 a.m. This order was carried out, and we recrossed into Knoxville.

October 9 – Left Knoxville by rail, going east, at 11 a.m. Forrest's Cavalry were raiding in the direction of Greenville. Near Bulls Gap, we left the cars and bivouacked for the night.

October 10 – Marching through Bulls Gap, we came upon our cavalry and mounted infantry fighting the enemy at Blue Springs. Our brigade under Ferrero was in the lead. A few minutes observations convinced Ferrero that our people were not doing justice to the occasion. The enemy was clinging to a wooded ravine with considerable pertinacity, and the cavalry were trying to shoot them out, but the shots did more damage to the treetops than to the enemy. Ferrero promised Burnside that with his permission he would take his brigade and clean out the ravine in twenty minutes. Receiving permission, the brigade was formed under cover of a ridge near the ravine occupied by the enemy. Ferrero dismounted and advanced alone nearly to the crest of the ridge, halted, faced about, glanced up and down the line, drew his sword and pointed to the top of the ridge. The line moved at once, up the ridge, over and down into the ravine held by the enemy. A storm of rifle balls greeted our men as they swept down into the timber. Without a halt, they delivered their well-directed volleys as they went, driving the rebels before them. In less than the promised twenty minutes, the enemy was in full retreat, and we bivouacked in the ravine for the night. Ferrero went through the bivouac personally, and said to the men, "They are killing beef over there. Go and take what you want. The butchers will object, but take it." Every man went and cut just what suited him. The butchers did object; they were hustled to one side as being of no account. We had one good feed of beef anyway, and Ferrero ordered that the rations should be issued in the regular way beside. The skirmish at Blue Spring was our first engagement in Tennessee. The casualties to the Union side were 100 killed, wounded, and missing; the Confederacy 216.

October 11 – The cavalry tried to bring the enemy to lag long enough to get a force in their rear. It was hoped that with a strong cavalry force in their rear, and the infantry in front, that Forrest's cavalry could be captured or destroyed. But the wily rebel had no intention of falling into this trap. Our cavalry kept up a running fight with him all day, and the infantry followed as fast as possible. The road was strewn with broken guns, demoralized wagons, some dead and wounded rebels and dead horses. At Greenville, many women and children came out to meet us and thanking God they had protectors. One old Negress, with uplifted hands and religious fervor, prayed, "De Lord bless de damn Yankees." Passing through Rhaetown, the infantry halted and left the cavalry to chase Forrest into Virginia. We were now eighty-eight miles east of Knoxville. Here as elsewhere throughout East Tennessee, the people had been skimmed of everything eatable. They, as well as the soldiers, were on short rations.

October 13 – After a day's rest in which many ladies came to visit the "Boys of the Blue Jackets," we began a march back to Knoxville, Four miles from Blue Springs, our bivouac was made beside the railroad. Trains met us here, but there were only cars enough to carry the sick and wounded and part of the baggage. Feeling sure of being transported on the cars, the mounted officers posted off their hostlers with their horses to Knoxville, so as to have them at hand on our arrival. But alas for human expectation, we did not get transportation, and the boys said, "The officers have to 'frog it' with the rest of us."

October 14 – Passed through Bulls Gap. Officers getting footsore and hungry like common soldiers. Anything eatable was a prize to be sought.

October 15 – Passed through Russellville and reached Morristown. Officers cursed their luck and declared they would never let their horses out of sight again. Privates smiled sardonically and murmured to themselves, "Ought to had better luck," "Know how it is," and like consolations.

October 17 – After waiting a day for cars, most of the troops got away on the cars for Knoxville in the morning. The 2nd Michigan alone had to wait until midnight. During the nearly two days wait here, I made two excursions for "grub" on my personal recognizance. Once I got a few biscuits, and the next time I took tea with an awkward but good-hearted family by the name of McKinney. This was the utmost that could be done to live off the country, and this could only be done in isolated cases. At daylight, we arrived at Knoxville and camped east of the city. Orders were promulgated that quarter rations only would be issued. To men who were doing hard service and had not been full-fed one day in five weeks, this order was depressing. The rations issued by the commissary was all there was to be had. Money could not buy where there was nothing to buy.

In the two or three days stay in this camp, the effect of these abbreviated rations told upon the spirits of the men. Hollow-eyed, cadaverous and spiritless, they hung over the smoky fires of green pine wood, the cold drizzling rain quenching the real fire and the fiery spirit alike, while they tried to make their smoked and dirty, tattered garments cover their nakedness and shield them from the rude blast. Keen eyes peered from sunken sockets over sad cheeks at the lean mules that ate their six musty ears of corn that they might have strength to bring more food for us across the mountains from Kentucky. Every one knew the mules were the salvation of us all. They were zealously guarded, yet many a mule ate five ears instead of six. Like hungry wolves, the men watched the keepers of the mules and an opportunity to steal an ear from their scanty feed was never lost. There had begun a universal, continual search for food.

October 20 – At 7 a.m., pursuant to orders, we marched westward down the valley toward Kingston—fourteen miles.

October 21 – A cold rain falling nearly all the time. Marched twelve miles and camped in a piece of timber two miles west of Lenoir's Station. Sometime previous to this, two comrades and myself had formed a partnership for the purpose of securing food. Whenever there was a chance, each was to forage for the whole. When we came into camp this time, one of my partners had captured a pig's head, very fat, but valuable under the circumstances. This was boiled, and we sat down to sup off it, without salt, pepper, bread, potatoes, or any other adjunct. The quivering fat of the pig's chops was all our store. Seated upon a little grass mound making the best of what we had, the old white headed Negro, Uncle Henry, the Captain's servant, approached and said, "Gentlemen, does you say you suffa?" I answered, "Very much. Are you hungry, Uncle Henry? Have a slice." And I cut off a generous slice of fat.

"O no, I'se not hungry. I'se just thinking how you ones would injoy a hardtack apiece, but for de Lod, I dunno whar you git im." With that he turned upon his heel, drew his hands from his pockets, down rattle three hardtack, and he walked away without looking back. It troubled us not a mite that probably he had stolen them from someone else to give to us. We ate them with our fat and questioned not.

October 22 – We moved down and crossed the river to London. The enemy were said to be in force near Philadelphia. The cavalry and mounted infantry were to make a reconnaissance in that direction while we were to remain here to support this if required. With ponchos pitched upon the bleak hills, we dragged out five rainy days and chilly nights, punching the clay hills into sticky mortar, and hugging empty stomachs to keep back the gnawing pain that held the place where food ought to be. One day when the sun shone out for an hour or two, I sought to forget for a time the general wretchedness by strolling upon the hills. From one of the higher elevations, the view was grand. Many miles to the north, standing out in bold relief against the sky is the Cumberland Mountain ranges, blue and cloud capped, rearing their lofty summits far above the lesser elevations that dwindle away to nothing as you approach the valley of the Holston, which lies spread out to view. The panorama reaches far toward the rising sun, embracing plantations, meadows, woodland, rivers, and undulating hills. The variegated autumnal tints of the forests, contrasted with the green of the pastureland, and the sober gray of the ripened corn, and the soft light reflected from the hilltops shading down to the dark shadows of deep ravines, adds beauty to the picture. Faraway to the southeast can be traced the black tops of the Smokey Mountains on the line of North Carolina. To the south, a range of hills obscures the view. The valley before us is a very pleasing garden spot, fit to be the abode of peaceful American citizens.

In the foreground of this picture are the low tents of the 9th and 23rd Army Corps, and just beyond is the beautiful current of the Tennessee River. The tread of armed men has marred this beautiful valley and wrought misery among its inhabitants for two years. And yet, human blood is to redden its hillsides and stain the current of its streams. Its habitations will be riddled with shot and destroyed with the torch. Its people will be impoverished and driven from their homes by the dread fiend of war.

October 28 – The reconnoitering party having returned, we were wakened before daylight without the aid of drum or bugle. Tents were struck without noise. No fires were allowed to be built. We noiselessly crossed the river, and the pontoon bridge was taken up. Rested near Lenoir's Station.

October 29 – Orders were issued to build winter quarters. The position assigned our brigade was a piece of timber over a mile east of Lenoir's Station. At once, the ground was laid out and each squad, mess, or other combination began work on its own account. Most of the men aimed to build huts of small logs, using their poncho tents for roofs. My two comrades and myself decided to have something a little more elegant. Remembering that there was a Methodist campground three miles east near the railroad, where a circle of board cabins circled around a central pavilion, we conceived the idea of removing some of this lumber to our camp to make us a winter home. A little search discovered one of these small flat cars that railroad men use to move rails and ties from point to point, as they are needed. With this somewhat rickety push car, we proceeded to the camp meeting ground, tore down a cabin, saving as many rusty nails as possible, loaded our car and pushed it up the heavy grade to a point opposite our camp. From here, our backs had to bear the burden. Two such trips secured us sufficient material for our "Palace." These labors banished in a measure the thoughts of our famishing condition and gave an interested look to the countenances of the men. The scene was a novel one. Axes and hatchets made the woods ring, and the huts began to rise, each company exactly in line as the busy workers strove to get into their new homes as soon as possible. The prospect of comfortable shelter was no mean item to people with our scanty clothing at this considerable city, but our individual cabin building was interrupted. As soon as camp was established, a court martial was opened at Lenoir's Station for the trial of a long list of minor offences that had been committed since we left Kentucky. Both of my companions were summoned as witnesses and left me to continue the building alone as I could find time with other camp duties. We had the sides up and roof on and used a blanket for a door.

One day as I worked on the inside arrangements, a sudden gleam of sunshine attracted my attention, and I turned just in time to see a black hand scatter half a dozen onions upon the floor. The hand was withdrawn, and the blanket fell to place. Soon after, I met Uncle Henry and told him someone had dropped some onions in my tent.

"Does you like onions?"

"Yes."

"Well, I guess you know what to do with 'im."

Another day, one of my partners dashed in with a basket and breathless, gasped out, "Have supper at six—I expect to be called as a witness every minute," and he was off like the wind. The basket contained two pounds crackers, two pounds cheese, one dozen eggs, two packages baking powder, and two bottles of whiskey. Of course, I cooked supper and used the basket to help make the necessary heat. When the boys came home, the following story of the basket was related while we ate such a repast as had not refreshed us for many a day. When the train came in from Knoxville, a sergeant got off the train with a basket for his colonel, and a bag of apples he had brought along as a speculation. Taking a position near the middle of one of the cars, he set the basket behind him under the car and proceeded to sell his apples three for a quarter. The temptation too great for my partner. Creeping under the car from the opposite side, he gently withdrew the basket and was congratulating himself that he had been unseen. When, alas for human hopes, he found that one fellow had seen all. Then and there a race began. Making a wide detour, hoping to outrun or lose the chase, the race was kept up until it seemed useless, so he sat down and waited for his pursuer. The fellow was willing to cry halves, but my shrewd partner got rid of him for a bottle of whiskey, some baking powder, and a half dozen eggs. The baking powder was of great value. If we could get flour, we had nothing but water to combine it with except sometimes we had salt, sometimes not.

During the week that winter quarters were building, our diet was corn on the ear, two ears a day to a man, and quarter rations of beef, supplemented by what individuals could forage. Forage in this sense means beg, borrow, steal, or carry away. My share of the week's forage was a little pork, a few hardtack, and an occasional extra ear of corn that lessened the feed of some skinny mule. The other partner had "cramped" a small pumpkin. Stewed pumpkin, browned corn, and beef without seasoning may not be very tasty, but it sure beats nothing.

Our house is completed, and I may as well describe it. In contrast to the log huts of the others, ours is of boards standing perpendicular, the cracks battened with thin strips. The roof is of tent cloth. A door is at the south and a window at the north end. Before entering, please stamp your feet on the platform and use the iron scraper at one side. As you enter, you are facing the window, to the sill of which is hung a swing table that can be lowered out of the way when not needed. Above the window is a narrow shelf where pens, ink, combs, brushes, and blacking could be stored. On the right, is a fireplace where the green wood fizzes and rolls up its black smoke. Above the fireplace is a broad shelf that could hold boxes and cans filled with flour, sugar, coffee, salt, pepper, soda, ginger, and mustard. There is also room for the plates, cups, knives, forks, spoons, and saucepans. On the right, is a cupboard parted in the middle to separate the bread from the meat. Over the mantle piece is a very small mirror that will show you if your hair is at the right curl. Beneath the cupboard is a well-filled wood box and behind it is a broom made from twigs lashed to a stick. At the end of the room opposite the fireplace are the bunks, one above the other. The blankets smoothly spread over the bed of pine boughs show the skill of the chambermaid. Facing about, you see on one side of the door three rifles with boxes, belts, and bayonets. On the other side of the door are haversacks and canteens. Lifting your eye to the ridgepole, you will see some ears of seed corn carefully suspended from it as if in anticipation of a shortage in the spring.

With what great hopes had we planned and arranged this domicile. Our captain remarked that we were well fixed to get married.

November 6 – Leaving our snug quarters, we were whisked away by train to Knoxville to anticipate some movement of the enemy. For three days, we bivouacked on a bleak rise of ground back of the railroad depot without shelter. The weather was cool, and rain fell enough to keep the ground wet and muddy. A train of supplies had arrived and plenty of provisions were issued, which helped wonderfully to withstand other hardships. The ingenuity of soldiers in getting what they desire was illustrated at this bivouac. A man living in a board house opened a trade with our men on molasses and cider. He had two barrels of cider. One was on top, the other in the room. While he was selling molasses or cider to suit purchasers, by the tin cup full, the boys got the extra barrel of cider out of doors and were rolling it away when his wife discovered them from the back door. Of course, the boys could not run away with a barrel of cider, so when he gave chase, they abandoned it. Telling them that he would give everyone a drink who would own to having a hand in stealing the barrel, he rolled it back and took it inside his temporary bar. The boys came up manfully and got their drinks. All went well until he found it necessary to tap the second barrel. He then found they had bored through the side of his shanty into the cider barrel, and drawn it all out through an elder stalk into their tin cups and canteens.

November 9 – Toward evening, we returned to our winter quarters. Four days we remained in camp in full enjoyment of good shelter. Some clothing had reached us, and the most needy were fixed a little more comfortable.

CHAPTER 16 – NOVEMBER 14, 1863

The day broke with lowering clouds and all the indications of a cold disagreeable storm. Directly after a scanty breakfast, all were surprised to hear the general call sound. Simultaneously came an order to issue rations. A cup of raw flour and a small piece of raw beef was hastily dumped into the haversack of each man and out from their comfortable quarters the men went, into the rain, which had begun to fall. Turning to the west, the deep mud in the road drove us to the railroad track.

Stumbling along as best we could, we breasted the sheets of cold rain that were dashed in our faces by fierce gusts of wind. Perseverance and time brought us to the river opposite London. Instead of crossing the river, the road leading down the right or north bank was taken. By this time, the clay hills had become water soaked a foot deep by the incessant rain, and the tired artillery horses could do little more than work themselves through the heavy clay mortar without the extra exertion of dragging the heavy guns after them. As usual, recourse was had to the soldiers to haul them. Men were stationed along the tugs and traces as close as they could walk, to pull while the drivers plied spur and quirt and cathes to the poor horses as they floundered through the slippery, sticky, treacherous mud.

Three miles below London, night overtook us beside some heavy timber. Drenched, chilled, and tired, we moved into the timber under the most stringent orders to build no fires, make no noise, and to remove none of our trappings. All day in the rain and mud, toiling without food, with raw flour and beef at night in the pitchy darkness of the woods, without fire, encumbered with wet garments and heavy trappings, the rain roaring down through the leafy trees, our hardy boys wore out the almost endless night.

November 15 – With the first faint light of day, a retreat was begun, which was continued slowly to Lenoir's Station, the enemy following without pressing us. At Lenoir's Station, our brigade was relieved from the front to get breakfast. Fires were quickly blazing, and the prospect of a cup of coffee brightened every countenance, and the sun shone out to cheer us. But alas for human hopes, before the water boiled, the enemy's bullets came pattering among us from across the river, and the picket line at our front came rushing in, broken, and in confusion.

Immediately came the order, "2nd Michigan, fall in," and we went to the front unified to restore the broken line. The line was reestablished without our firing a gun. Expecting that whatever had driven the others from their post would soon give us trouble, we stood to arms with knapsacks on the remainder of the day nearly through the night. no further demonstration was made on that part of the line. The outpost had evidently been frightened by a scouting party.

November 16 – Before daylight, we were called in, and at Lenoir's Station, our brigade had been detailed as rear guard. A train with supplies for the 23rd Corps had arrived at Lenoir's Station, but the drivers being frightened to find themselves so close to the enemy, cut loose their teams and fled, leaving the loaded wagons in the road. These were fired, and otherwise damaged, as much as the limit of time would permit, but most of the supplies went to comfort the rebels instead of the famishing boys they were intended to cheer. With the first light of day, the enemy began to push the rear guard of our army. It had been decided to move the army to Knoxville, and it was designed that the rear guard should delay the enemy's advance while the artillery and baggage trains were got as far on the way to Knoxville as the condition of the roads would permit. This difficult task had been assigned to our brigade, 3rd Brigade, 1st Division, 9th Crops, whose decimated ranks would hardly equal a full regiment, under command of Colonel Wm. Humphrey. The enemy was bringing up an army of 30,000 men on two roads that formed a junction about two miles from Campbell's Station. We were on one of the roads, the other was unguarded. The problem was this, to keep the enemy from advancing fast enough to capture any of our baggage or artillery, and at the same time, reach Kingston junction before the enemy's other column arrived there and cut us off from our army. With this object in view, our tactics were to fight and fall back. With a simple piece of artillery to annoy the enemy at long range, and our line deployed on either side of the road, we made a stand upon the first high ground and watched the enemy advance with wide extended lines to overlap ours, his flank thrown well forward as if intent upon closing in upon us, if opportunity afforded.

Waiting for him until he was within range, we opened a deadly fire that checked his advance, and then we retreated, to repeat the same move on the next rise of ground. At one place, we descended a long hill, crossed a stream, Turkey Creek, and ascended the opposite hill. Our regiment was not permitted to pause long enough to fill our canteens with needed water at this stream. Colonel Comstock of the 17th held the creek as long as possible, and in consequence, lost thirty or forty men before ascending the opposite hill (Seven killed, nineteen wounded, ten missing).

At last, we were at the top of the hill that leads down to Kingston Junction. A triangular piece of timber lies between the two roads. As we entered this timber, the enemy made a dash at us and followed closely down through the timber. At last, we reached the junction. Out upon the open plain along the Kingston Road, was the other rebel column in dense masses pressing on almost within rifle shot. The road makes a sharp turn here and there, and our line must swing round to cover this road. Something must be done quickly, or we will be overwhelmed in flank and rear. Two men from each company were detailed to make a dash into the woods, while the whole rear guard set up such a hurrah as never came from so few throats. As they dashed forward, the rebels recoiled and our line swung round just in time to present a square front to the united foe. The detailed men were abandoned to their own chances. Of the two detailed from my company, one returned, the other did not.

A succession of hills rise one above the other to the east of Campbell's Station. Upon these hills, Burnside was arranging his line of battle to make a stand against the enemy. The rear guard continued its retreat to Campbell's Station, halted at the base of the hills, and held the enemy in check until after noon when it was relieved and went a little to the rear. Halting in a little ravine beside a little creek over which flew the shots of friend and foe, we cooked and ate the raw beef and flour that was given us two and a half days before. During the last stand, our color bearer was shot while resting his elbow upon my knee. The loss during the day was Union killed sixty, wounded 340, more than two-thirds of which was sustained by the rear guard. The rebels, 570 killed and wounded.

Having broken our long fast, we were moved up the hill nearly a mile, all the time under fire of the enemy's artillery, the shots tearing up the ground about us and the bursting shells scattering their fragments through our ranks. With a coolness amounting to indifference, the brigade marched up the hill and took position to support a battery on the second line. This duty allowed our boys to rest their tired limbs while they lay on the damp ground for any call. Night closed the battle, and the retreat was continued. The clouds were again lowering, and rain fell at intervals. The wagons that had preceded the army had cut the roads into a horrible condition. Through the black night—to us the third without sleep—we floundered on through the mire, resolutely forcing back the desire to join the groups of stragglers that here and there had piled up the fence rails and built fires to warm and dry themselves. The rays of light from these fires revealed exhausted men in every attitude. Some lay prone in the mud, some partly borne up by a fence rail and partly in the mud, some sitting in the mud, and reclining against a stump or tree. Cavalrymen fell from their horses, from inability longer to keep their seats in the saddle, into the slush of the road. The whole scene was one to be remembered by all who participated. Throughout the miserable night, the only sounds heard were the slump, slump, slump of the soldiers' feet in the pudding-like mud as they trudged on through the darkness.

November 17 – Completely worn out, little caring whether the world revolved any longer or not, our brigade halted on the ridge near an incomplete redoubt in the northwestern outskirt of Knoxville. At the command "Rest," the men dropped their knapsacks from their shoulders, and without taking a single step, lay down in ranks as they had stood, fell into a dead sleep, and forgot for a time their hardships. After daylight, the hour I did not know and had no care to know, I was roughly awakened and ordered to issue forty rounds of ball cartridges to the men of my company. Alone, I went to the quartermaster and got two heavy boxes of ammunition, tugged them, one at a time to the head of my company. The men lay upon their backs in line as they had dropped down when ordered to rest. The pale, haggard faces that were turned toward the sky will remain a vivid picture on my memory as long as my own faculties are intact. I attempted to arouse the men to give them the ammunition, but the task was in vain. No efforts of mine would induce them to rise or even to awake. Giving up the attempt to rouse them, I piled the cartridges on each man's breast, and let him sleep on. It is needless to add that when they did awake, each cartridge was cared for as a treasure. When my own exertions in attending to this duty had sufficiently dispelled the semi-lethargy that loaded me down, I began to take an interest in the surroundings. On the brow of the ridge, was a line of men vigorously wielding spades and pickaxes. Their faces exhibited all the colors from the purest Caucasian to the blackest African, and their dress revealed all the conditions of human society from the basest slave to men of wealth and affluence. Some worked cheerfully as if they were in favor of the work. Others showed by their sour looks that there was a power behind them that compelled the work against their will. These men were picked up in the streets of Knoxville as they appeared on the streets and marched out to throw up breastworks for us. No excuses were taken from white or black, high or low. If found on the streets, they had to come. By noon or a little after, the infantry had all arrived at Knoxville. The various commands were assigned their various positions around the city and lost no time in getting to their places. The 2nd Michigan took possession of a position of the rifle pit that had been thrown up immediately to the right of the redoubt near which we had slept. Some of our brigade were in the redoubt and some were to our right. Every man gave his best efforts to perfect the defenses in front of his own command, and men were put to work to make a fort out of the redoubt. While this was going on, Colonel Sanders with his cavalry was holding the enemy in check about a mile from our position. The music of his guns encouraged us to strive on with the defensive works. Burnside visited each command, encouraging the men and told them they must hold Knoxville, or go to Richmond as prisoners of war.

November 18 – Work continued on the defenses. Colonel Sanders still held the enemy at bay. He kept up the fight until half past two when he was killed. He had accomplished what he was set to do—make time for us to build defenses—but it cost his life. The fort near us was named in honor of his heroic sacrifice. The redoubt was fast growing into a fort, and after this day, was known as Fort Sanders. Fort Sanders occupied the brow of the ridge where the ridge angled so that it sloped away to the west and north, thus making the most important corner of the fort its northwest angle. The rifle pit, following the brow of the ridge to the east from Fort Sanders, was our position. In front of us was a long graded slope down to the railroad track. This slope was covered with a straggling growth of pines, averaging about one foot through. These were cut down and placed above the earth of the rifle pits on crossed stakes, leaving a crack between the logs and dirt of about four inches. Behind the works, it was constructed so we could fire through the crack without very much exposure to the shots of the enemy. Directly in our front and across the railroad was a considerable portion of the poorer dwellings of Knoxville. To the left of these was an open level plain extending a quarter of a mile to a stretch of heavy timber. By the time that night closed in, our pickets were in the railroad cut, angled out across the plain in gopher holes and in among the dwellings on the outskirts. With the enemy pressing right up under our noses, no thought was given to our lack of food and shelter. Every man taxed his brain and energies to improve the defenses. Although everyone realized that the situation was a grave one, there was no depression of spirit. Jokes were perpetrated and laughed at as heartily as upon any ordinary occasion.

November 19 – Heavy skirmishing at various points about the city indicated that Longstreet was rapidly closing in to invest the town. The inhabitants left their homes in the outskirts of the city, abandoning everything and sought shelter elsewhere. The soldiers appropriated whatever of eatables or bedding they left. Stray rifle balls were frequently dropping inside our works. Sergeant Bell of Company C, while receiving directions from his captain, was struck on his upper lip, the ball penetrating his head opposite his ear. He was taken away to the hospital to suffer for a time. Like casualties were frequent, but war takes little account of individual mishaps when there are so many great slaughters. Everyone was intensely alert for any emergency. Hunger was gnawing at our vitals, and the supply of provisions was distressingly scarce. Small pieces of plug tobacco were issued with the small allowance of other rations. A party of men were detailed and supplied with balls of cotton and camphor to fire buildings in our front in case of a night attack. This evening there was a slight demonstration at the outpost, and several houses were fired. These at intervals along the entire north front of our lines lighted up the surroundings so that any movements could be seen. The drum corps of each regiment beat the tattoo twice or three times to convey an idea of a much larger force than we had.

November 20 – Light infantry firing, but little artillery firing, indicated that the enemy was perfecting preparations for some kind of action. When not on other duty, the men kept on improving the defenses. The enemy's pickets had occupied James Armstrong's house just under the hill to the west of Fort Sanders. In the evening, the 17th Michigan was ordered to dislodge the enemy from the house and burn it. They successfully accomplished the task in a gallant manner without the loss of a man, but when returning, the light from the burning building revealed them to the enemy who opened upon them with grape shot and killed one lieutenant and wounded three men.

An incident showing the ability of our soldiers to keep their wits about them in the most trying moments comes out of this sortie. The enemy in the Armstrong house was just preparing for supper when our men surprised them. A batch of biscuits hot from the fire was upon the table. When the 17th were safe inside the works, the pockets of a captain were observed to bulge out enormously. Being questioned about them, he drew forth the biscuits, still warm, with the remark, "It made me feel so bad to see those biscuits in danger of being burned." This same evening, twelve other houses were burned along the line by the "Camphor Party," as we called those whose duty it was to set fire to buildings.

November 21 – Heavy rain fell in the morning. The occasional whistle of a rifle ball was all that broke the monotony of the siege. We starved, waited, and watched.

November 22 – The light of morning revealed a line of rebel earthworks across the open plain to the northwest. From a redoubt of considerable pretentions in the center, the enemy easily sent musket balls inside our works, greatly annoying us. Rations were reduced to one cup of coffee, one slice of fresh pork and a slice of bread three inches square, the bread of bran and sweepings of the mill floors, for a day. Under this feed and pressure of duty, the men remained cheerful and determined. There was not a murmur of repine.

Under cover of darkness this evening, Ed Curtis and myself went down among the deserted houses in search of eatables. Not finding anything in the more easily accessible houses, we went to the outpost, which was in one street, with the enemy's outpost in the next street. Feeling our way into a dark house next to the enemy's line, we found quilts in the bedroom and a small sack of potatoes, part of a sack of flour, and a bottle of molasses in the pantry. Taking the eatables, a quilt and chair each, we returned to our post without raising an alarm.

November 23 – The enemy was suspiciously quiet all day. In the evening, the outposts were alarmed by signs of a night attack. Instantly, without confusion, our troops were in readiness and fifty houses leaped into flame along the front. This included the railroad depot. The light revealed the enemy drawn up ready for attack. The wall of fire and the great light rendered their intended move impracticable, and they retired into the shadow. As the buildings burned, shells began to explode in many of them where they had been concealed by the rebels before they were last driven from the place. The explosion of shells and popping of cartridges made all the noise of a battle. From our security, we peered through the crack in our earthworks at the fireworks, while the great sheet of tin on the roof of the depot flourished in the heat with a noise like thunder.

November 24 – Day broke with a heavy mist hanging over hill and valley, completely shutting out all view of the enemy's lines from us. As soon as it was light enough to see, the 2nd Michigan was ordered to "fall in." Without the usual cup of coffee, the line was formed, and the roll was called. In the absence of the Sergeant-Major, I collected the reports of the orderly sergeants of the various companies and turned over to the adjutant, a report of 160 men present for duty in the regiment. Immediately, this little band moved down the slope into the railroad cut. Halting here, we were informed that our destination was to capture the center redoubt out on the open plain. There was no remonstrance against the undertaking, but the general expression was that one-half of us would not return alive. The signal to charge was to be one gun from Fort Sanders. The signal to retreat two guns.

While we waited the signal, the utmost freedom was given the discussion of the chances of our undertaking. There was no thought of hesitation. There was no blanching. And when the signal came, with set teeth, the men sprang up the bank as one man. A slight breeze had cleared the mist, and the sun poured its glad rays upon many a brave boy for the last time.

Wheeling slightly to the left to squarely face the objective point, our line moved rapidly forward. The watchful foe, discovering us, poured in a deadly volley. Several fell at the first fire. The guns of Fort Sanders opened a point-blank fire on the work over our heads. With wild mad shouts, our line dashed on into the vortex of their curved line where they poured on us a murderous fire from both flanks as well as in front. Bullets screamed and whistled through the air from all directions. There seemed to be lead enough in the air to almost shut out the light of the sun. As we came within range, the guns of Fort Sanders ceased firing, leaving the fight all to us.

Coming upon the redoubt, our right swung forward so that they could fire directly into the end of the redoubt. At the first volley, the enemy broke and fled back to their reserve in the woods close in their rear. Now having full possession of the earthwork, we took shelter on its outer face and fired over the top at the foe in the timber. I had seen Adjutant Noble and Lieutenant Galpin fall before we reached the work. Lieutenant Gulver had also been killed. Men had been falling all the way and were dropping every second now. While aiming over the work, a ball struck me in the right breast, passed through the lung and out at the back near the spine. Major Byington who was in command, came and said to me, "You are badly hurt," passed on and had not gone ten feet when he was hit almost simultaneously in the leg and in the side. Raising upon his knees he called, "Pass the word to Captain Ruckle to take command and tell him for God's sake to get the boys out of this." Just at this moment, the two guns belched forth the signal to retreat. Those who were able got away. The enemy now advanced to reoccupy the work, and the infantry around Fort Sanders commenced firing to cover the retreat of our men. As the rebels approached us, they began calling to one another to bayonet the wounded. A swarthy rebel major with long black whiskers and a cocked navy revolver in each hand, dashed to the front and roared out, "Who says bayonet the wounded? Show me the son of a bitch." There was no more such talk.

Reaching over the work, some rebels dragged me inside by the collar. I was semi-conscious by this time, but I remember that a rebel sergeant took my watch and gave it to one of my comrades with instructions to send it to my friends if he got out alive. I also remember the big rebel major ordered four men to place me on a stretcher and march in step without flinching back to the woods. This order he enforced at the muzzle of his two revolvers, although the air was full of shrieking leaden missiles of both friend and foe.

This action did not last half an hour, yet in that time, 160 men had driven 200 men from an entrenchment, they being supported by a whole brigade not eighty rods away.

Eighty-six out of the 160 were killed and wounded. No advantage had been gained. It was impossible to hold the work without moving a large force outside our entrenchments. This was apparent to most everyone before the work was taken.

In view of all the conditions, the person or persons ordering this move were pronounced guilty of official murder by all who witnessed it.

CHAPTER 17 – NOVEMBER 24, 1863

The stretcher-bearers deposited me on the grass near a little brook in the woods. I must have been dazed for I do not recollect seeing any other wounded men or many others near me. A rebel straggler came and asked what he could do for me. I asked for water to quench the burning thirst that was upon me. I drank his small cups of water about as fast as he could bring them from the brook. He stayed with me for some time. Finally telling me he must go, he asked if I had a jack-knife, pocket book, or any green backs to sell. I gave him the old knife I had in my pocket for his kindness. This he refused to accept without pay and thrust two Confederate dollars in my bloody pocket. After a time, I was loaded into an ambulance and driven by a wood road, the longest journey of my life. The distance was not more than two or three miles, yet it seemed everlasting. Every root or stone the wheel struck hurt me terribly. The driver was as careful as he could be, yet every jolt seemed as if it would take my life away. At last, we reached Middle Brook paper mill situated upon a creek of the same name in the woods. In this mill, the rebels had improvised a hospital. Around the door stood a group of surly looking rebel soldiers. The ambulance driver called for help to get me out of the ambulance. No one moved or said a word. He begged for help with like results. He cursed and swore, but got no help or words. At last in desperation, he seized me by the heels, dragged me out far enough to get my middle on his hip under his arm. In this manner, he carried me as he might have carried a bag of wheat into the mill and let me drop upon the floor. For a time, I lost entire consciousness. After a time, an attendant dragged me around upon some paper clippings and informed me that was to be my bed. Soon after, I discovered Major Byington was lying with his feet to mine and Sergeant Dix lying at my side with some other members of my regiment and rebels filling the room. Without blankets or cover of any kind, we lay upon this handful of litter in the fireless old mill, with a cold raw autumn storm coming on. A dozen or more of the 2nd Michigan were there and perhaps fifty wounded rebels. Amid the groans and howls of the Confederates, we "Yanks" bore out the almost endless night, enduring our pain and discomfort in silence. We received absolutely no attention, except occasionally, a mouthful of water when we could intercept a nurse who had been awakened by the importunate howling of some Confederate sufferer. At such times, these rebel nurses would say, "Why don't you-ons howl like our men?"

November 25 – Through the open doorways, we could see the cold drizzling rain come down outside, while our feverish bodies were drying our blood-soaked garments until they gouged our wounds at the slightest move. What a longing for plenty of water to cool the parched lip. What a desire to be freed from the stiff and cruel clothes. What a pleasure a bath would give and how grateful a mouthful of food would be, no one can guess who has not had the experience.

During the day, the stalwart rebel discovered I had a good pair of boots and proposed to buy them off me with Confederate money, saying, "They will do you no good." Thinking he would take them anyway, I proposed to trade them to him for a blanket and an old pair of shoes. He accepted my offer and brought an old blanket with many holes and a worn out pair of shoes many sizes too small for me. The blanket did me some good. Just at dark, a nurse brought us two soggy biscuits and a thin slice of fresh pork, the only food we had taken in forty-eight hours. As soon as it was fairly dark, the nurses all went to bed and left everyone to care for themselves. The suffering from cold and thirst and stiffening wounds was terrible. During the night, a fat rebel captain who had been brought in wounded, vigorously denounced the Confederacy, winding up his bitter tirade with "A cause that gives so poor care and little attention to its wounded is not worth fighting for." A citizen of Atlanta, Georgia, with one leg drawn up, was caring for his son near me. Upon my complaint that it hurt me when his short leg came down, he said, with tears in his eyes, "I will be as careful as I can, but my poor boy is dying. If there is anything I can do for you, don't be afraid to ask me." The poor boy did die, and as soon as his father left, him the rascally rebel nurses woke up and went through his pockets.

This night and all day November 26, we were left without care. Such a long night and such a distressful day are never to be forgotten.

November 27 – A black-haired, black-eyed, dark-complexioned Confederate soldier came to me and asked in kindly tones what was the nature of my wound. After examination, he informed me he did not belong to the hospital, but would try to make me more comfortable. Procuring a hand basin and bandages, he washed my wounds carefully and was about to dress them when the basin and bandage were snatched away with, "We cannot attend to the amputation cases alone without caring for others."

However, he seized the end of the roll of bandage and tore off enough to pass around my body, between my stiff clothes and the wounds. The litter of paper was taken out, the floor swept and fresh straw was put on the floor for us to lie upon. Dr. Burton, the rebel surgeon in charge, came and looked at me and started to pass on. I hailed him with, "You old rebel, what do you think about it?"

"Hello Yank! You're going to live."

"Who said I wasn't?"

"Well, what's the matter with you, anyway?"

"I should think, after a man had been under your care for a week or two you would try to find out."

His head dropped upon his breast, and he replied, "The lamentable fact is, we have almost nothing to do with." After a little more talk, he produced two morphine powders from his vest pocket and gave them to me to take when I could not rest without them. He ordered the nurses to give me all the whiskey I would take and left me. And that was all the care of attention I received at the hands of the rebels.

November 28 – What a long night of fever and thirst. Federal and Confederates were moaning for water throughout the night. Nurses taking their rest undisturbed by the distress of friend or foe. I rested easier during the day under the influence of morphine and whiskey. Toward evening, I saw large masses of troops moving toward the Confederate right. I informed Major Byington, and he questioned the surgeon about it. They said they were preparing to assault the Union fort (Fort Sanders). What hopes and fears for our comrades in Knoxville kept our pain and suffering company through another long night.

November 29 – About three o'clock in the morning, there suddenly burst upon our ears the sound of rapid artillery firing and musketry in the direction of Knoxville. The simple inquiry, "Are you awake," were all the words that passed between us wounded Union prisoners, but pain and discomfort were forgotten in the anxiety for the outcome. The heavy detonations following one another in quick succession and the ceaseless heavy roll of musketry assured us that the action was fierce and bloody. How we longed to be with our own brave comrades, and how we waited for day to reveal to us the results. Long before it was light, the firing slackened and finally ceased altogether. Silently, we waited for the surgeon to appear, and when he came, the lack of jubilation in his manner told us plainly enough that Fort Sanders had not been taken.

Major Byington asked, "What success, Doctor?"

"Oh! We planted seven stands of colors on the fort, but we failed to take it."

"How many did you lose?"

"Oh, not many."

The major, with great effort, raised himself on one elbow and said, "You never planted seven stands of colors on Fort Sanders without losing a thousand men."

We afterward learned they lost over 1,200, over 400 dead, and our loss was three killed and three wounded.

After noon, two officers from General Longstreet's staff came with paroles for us to sign. Toward evening, the Union wounded were placed in ambulances and sent to our own lines. Major Byington and myself were placed in the same wagon, the driver was changed at the picket line, and we were taken to the courthouse in Knoxville. What joy was ours to be once more under the protection of the glorious old stars and stripes with our own faithful comrades to care for us. My first inquiry was, "Have you got anything to eat?"

The nurse laughed at my eagerness and said, "Just as soon as I can get these stinking clothes off you." When my clothes were taken off, they were so rigid with dried blood that they stood up straight and firm as sheet iron. In a few minutes, they brought me two large crackers and a half pint of beef tea. My next tribulation was to know if that was all I could have. The nurse laughed and answered, "That's all tonight. You'll pull through without doubt."

What a release from trouble and what a sweet sleep I had that night.

December 1 – My wounds were thoroughly cleansed and properly dressed for the first time.

Plenty of good food was given us, and we rested peacefully and comfortably as we could in our condition. A new ward was formed on the third floor over stores adjoining the courthouse, and I was the first man moved into the ward. There were seventeen men put into this ward, mostly wounded men. As there was no actual disease upon these hardy sons of war, they were rather a cheerful lot. There was no great amount of despondency, and their philosophical acceptance of the wounds they had received left room for many pleasantries and some practical jokes.

Although we were under substantial roof, this was in other respects a field hospital. There were cots and soldier nurses. The two meals a day allowed us were prepared by soldier cooks in very simple style, and the food was very similar and as scanty as that furnished the well soldiers.

My cot was in the corner of the room by the side of a window. The broad window ledge furnished me table and storage room for any little trifles I had. The first night in this ward, my next neighbor was very restless, and the next morning, they carried the poor fellow out dead. His stay was so short, I did not learn his name. His place was filled the same day with a patient who was in the delirium of fever. His ravings were wild and incoherent. Two days finished his career, and he was "mustered out."

Cots in this hospital could not be idle. My next "pard" was a young

Pennsylvanian whose general appearance indicated that he had little disease, but some homesickness, yet not enough of that to destroy his appetite. He fidgeted continually, often searched his haversack and usually sat sidewise on the edge of his cot. The other boys goaded him and stuffed him with all sorts of nonsense. Among other things, they told him, "That fellow in the corner had killed every man that had lain next to him since the ward was started." I asked him why he searched his haversack so much. He said he was looking for his poke-bag (salt-bag). Being asked why he sat sidewise, he said, "I've got a bealing on the hinch" or boil on the haunch. The boys made him so afraid of me that he asked the surgeon to let him go to his regiment. He went.

Hospital life for two and a half months was uneventful, save the little incidents that show the spirit of the men who were waiting for their wounds to heal. The deaths in our ward during that time numbered four, the two previously mentioned, a poor fellow who was shot in the bowels, who stood upon his feet, talked some time in a strong full voice, sank back upon his coat and was dead, and Michael Livingston of the 2nd Michigan, who was shot through both lungs, who dressed and went down two pair of stairs, came back and died. Dr. Vickery of the 2nd Michigan bought chickens with the regimental funds and furnished the wounded each with a cup of chicken broth, every day, as long as the money lasted. Personally, I received from members of my company, some delicacies that the boys sorely needed for themselves. Sergeant Sexton brought me some applesauce. Corporal Curtis brought biscuit and pigs foot. Sergeant Lovett brought me a chicken, which the cook took charge of, dealt it out a little at a time. And John Schwartz, the cook, one day when our Scotch doctor in a freak had ordered me an extravagant meal, no article of which could be procured, brought me the nicest sweet Johnnycake I ever tasted. These helps to our hospital fare kept us up wonderfully.

The personnel of this ward was in part made up of Sergeants Dix, Stinebeck, Taylor, Thurlby, all wounded in the leg; Corporals Fuller and Lee, also wounded in the leg; Private Percy, leg wound. All of these were of the 2nd Michigan. Pat Kehoe of the 29th New York, wounded in neck and Malcolm Sinclair of the same regiment, wounded in mouth.

None of these were sick, and the spirit of mischief was prevalent among them. Corporal Fuller was very much inclined to be despondent, and the others, in genuine sympathy for him, conspired to rouse him by many a prank, played off in such a way as to arouse in him a fierce anger. He always was more cheerful and hopeful after his anger had cooled.

General Sherman came with two divisions from Chattanooga and raised the siege December 5, 1863. Longstreet moved up the valley toward Virginia.

Thirty days after I was wounded, I went upon the streets and continued to do so as long as I stayed in Knoxville. Dr. Moor of Massachusetts superseded our Scotch surgeon. Moor was a young good-natured Christian gentleman. He laughed at our pranks even when they were against the rules, saying, "The poor boys must have amusement.'

We roamed the streets at will, visited other hospitals, and played cards to pass tedious time away.

January 8 – There was a light snow upon the ground. The rebel spy, Captain E.S. Dodd of the 8th Texas Cavalry, was to be hung. Most of the inmates of our ward witnessed the execution. He was taken from the county jail in an open wagon, seated upon his coffin, and surrounded by guards with fixed bayonets, marched through the city to a little hill near the railroad where several Union men had been hung while the rebels were in possession of the town.

He marched firmly up the steps of the gallows, the order for the execution was read, the rope placed around his neck and the drop fell. He, being a heavy man, his weight broke the rope, and he went to the ground alive. Again, he was assisted to the gallows, the rope was doubled, and this time the fate of a spy was his.

Troops were frequently moving through the city, sometimes in one direction and sometimes in another. Reports often reached us that the enemy was advancing. At such times, we had various plans on foot to build a raft and quietly float down the river to the Union. This, however, never became necessary.

January 28, 1864 – Lieutenant Fletcher took all the wounded 2nd Michigan men of our ward to Blains Crossroads to see the regiment. It was a pleasant day, and we had a glorious ride and happy time with our comrades. In the meantime, the 2nd had veteranized and were to go home on furlough. Dix, Thurlby, Taylor, and myself were given furloughs to go home with them and report to St. Mary's hospital at Detroit, Michigan.

February 12 – We left Knoxville, homeward bound, by the way of Chattanooga, Tennessee, Bridgeport and Stevenson, Alabama, Nashville, Tennessee, Louisville, Kentucky, Indianapolis, Indiana. Arrived in Adrian, February 19, 1864.

To be once more amid the scenes of childhood and among old friends was a wonderful satisfaction. The comforts of home and plenty to eat was a blessing.

For a few days, scores of old friends called to visit and extended invitations to receptions. For a month this continued.

March 16 – Was admitted to St. Mary's Hospital.

Mrs. Jane Brent introduced herself by a letter from Colonel Poe. She took a great interest in members of the 2nd Michigan and did them numbers of great favors during their stay in Detroit. No effort was too great for her to put forth for them. Those who received her favors unite in saying, *God Bless Her.*

After some vexation and delay, I received my discharge from the army on May 31, but my papers were dated back to May 24.

Having been released from the restraints and dependence of military rule, a new mode of life is opened up and ways and means of livelihood have to be considered.

Within a week, I had been transferred from the military army to the army of working men.

CONCLUSION – OCTOBER 2013

My great grandfather, Harmon Camburn, was released from his service in the Union army one year before the war ended. In that final year of civil strife, soldiers and civilians continue to die in the bloodiest and deadliest war ever fought by citizens of the United States.

The war lasted from April 12, 1861, to April 9, 1865, when Confederate leader General Robert E. Lee surrendered his troops at the Appomattox Court House to Union leader General Ulysses S. Grant. Most historical records state that 625,000 U.S. citizens were causalities of this war. A recent study places the number closer to 800,000. No matter which number is used, the War Between the States is still the costliest one in the payment of human lives during its a four-year duration. From the perspective of 2013, it's easy to shake our heads at the shame of citizens fighting against one another for causes blurred and confused with economics, bigotry, politics, and power. Let us hope we learned valuable lessons from the priceless legacy left.

Comparing the death counts for other wars with the numbers from the Civil War, paints a ghastly picture of the toll on our citizenry.

American Revolution (1775-1783) - 25,000 casualties
War of 1812 (1812-1815) - 20,000
Civil War (1861-1865) - 625,000
World War I (1917-1918) - 117,000
World War II (1941-1945) - 405,000
Korean War (1950-1953) - 37,000
Vietnam (1955-1975) - 58,000

When I look at the number of lives lost, I am astounded that my great grandfather's severe wound—untreated for days—didn't make him a statistic in the fatality charts.

If he hadn't survived, this record of the realities of war from the perspective of a soldier would not have been written. If he'd succumbed in the war, he never would have married Eliza Holdridge in 1864. If they hadn't married, my grandfather wouldn't have been born and become the Reverend Arthur Thomas Camburn. And the good reverend wouldn't have married my grandmother Anna Mary Sweet who then gave birth to my father.

It's a miracle. When I think about the value of one person's life, I am grateful he lived so I now can share this war hero's story so many years later.

OBITUARIES OF HARMON CAMBURN

My grandfather is the Rev. Arthur T. Camburn mentioned in the obituaries. My father, Burtis Harmon Camburn, was born in 1904 and is one of the "five" grandchildren cited in the obituaries.

Adrian newspaper (March 23, 1906)
H. Camburn
"Death of a prominent citizen of this city"
"Was an old soldier with a good record"
"had been ill for the past two years"

On Thursday at 4:30 a.m. at his home, 61 Dennis Street, occurred the death of Harmon Camburn, the well known mail clerk of the Lake Shore road and veteran of the civil war. Death was due to a complication of diseases, resulting, no doubt, from a wound received during the war. The funeral will be conducted from the residence Saturday afternoon at 2 o'clock, with burial at Oakwood cemetery.

Mr. Camburn had been ill for about a year and a half from a complication of the heart and lungs. Last winter he had a severe attack of neuritis, and was unable to go south as he had planned. This winter, however, he went to Mt. Dora, Fla., but the trip did him no good. He returned home two weeks ago Wednesday and since that time had been gradually breaking down until the end came this morning at the hour mentioned.

Mr. Camburn was born in Franklin township, this county, February 4, 1842, and always lived in the county with the exception of three years spent in the civil war. He enlisted in Co. D, Second Michigan infantry, and engaged in many battles, including both engagements at Bull Run. He was wounded in the siege of Knoxville, where the Second regiment was sent against Fort Saunders after two other regiments had been driven back. His injury was a bullet through his left lung, which had caused much suffering ever since.

After the war he came back to Lenawee county and spent one year on the farm. He was married September 14, 1864, to Eliza Holdridge, of Raisin, and they came to Adrian, where they had since resided. For about 30 years Mr. Camburn had been a mail clerk in the service of the Lake Shore road. Most of the time his run was between Cleveland and Chicago, but for the past ten years or such a matter he had been running between Adrian and Monroe. Because of poor health he had to give up his work a year ago last November, and sent in his resignation last June. Besides the wife he leaves the following children: George M. Camburn, of Adrian; Rev. Arthur T. Camburn of Azalia, Monroe county; Miss Edith M. Camburn, a teacher in the public schools, and Mrs. Addie M. Church, also of Adrian. In addition there are five grandchildren and one brother and one sister, T. M. Camburn of Tecumseh, and Mrs. Martha Hollenbeck, of Alma. The deceased as a member of the G. A. R., and also of Adrian Lodge, No. 19, F. and A. M.

March 22, 1906
DIED AT HIS HOME EARLY THURSDAY MORNING
HE WAS A BRAVE SOLDIER DURING THE TRYING DAYS OF THE CIVIL WAR
RECEIVING A WOUND IN DEFENSE OF HIS COUNTRY THAT FINALLY CAUSE HIS DEATH AFTER TERRIBLE SUFFERING

As a Citizen, Soldier and Friend He Was Ever Faithful.

The very sad passing of Harmon Camburn occurred Thursday morning at 4 o'clock after an illness of nearly two years. The immediate cause of his death was lung and heart trouble, but ever since the war, during which he was badly wounded, he has not been strong. With Mrs. Camburn he went to Florida last November where it was hoped his condition would be bettered, but he became worse and they returned to Adrian on March 7. Since that time he has been growing steadily worse. He has suffered a great deal during his last illness, but has always maintained a spirit of fortitude and always had a kind, patient word for his immediate family, friends and members of the Grand Army, who often visited him. Very quiet and unassuming he has led an exemplary life, been a very good citizen, a home loving man and a kind father and husband.

His demise will be sorely felt by all those who ever had occasion to come in contact with his personality and his life will remain as a memorial of good deeds and acts, originating from honest convictions.

The funeral will be held at the home, 61 Dennis street, at 2 o'clock Saturday afternoon. Interment at Oakwood.

Harmon Camburn was the youngest child of the sixteen children of Mr. and Mrs. Wm. Camburn. He was born on Feb. 4, 1842, being sixty-four years old at the time of his death, at his father's farm in Franklin township this county. He attended school in Franklin when a young boy and later at the Graham school north of the city. This school was also known as Aunt Laura Haviland's as she was once the teacher in that district. He entered the war of the rebellion in the year 1861, enlisting in the volunteer service as Sergeant in Co. D., Second Michigan Infantry.

The late General William Humphrey was the captain of the company at the time of Mr. Camburn's enlistment. He was engaged in many of the larger battles of the war, including First and Second Bull Run, Fredericksburg under Burnside and Malvern Hill under General George B. McClellan. He was shot through the right lung at the siege of Knoxville and was there taken prisoner, remaining in the hands of the Confederates for six days when he was exchanged with a number of other prisoners and sent to the Union hospital in Detroit.

He had thus been in the service about three years. The wound was of a very serious nature, and had it not been for his strong constitution and constant care he would undoubtedly have succumbed to it.

After he recovered he came back to Adrian and was engaged in several branches of work until 1874 when he went into the mail service, where he remained for thirty years. For six years he made what was called the long run, that between Cleveland and Chicago. Altogether he made different runs between Cleveland and Chicago for twenty years. The last ten years of his service he ran between Monroe and Adrian. He had to give this up last year because of a general breakdown, and since that time has been endeavoring to recover his health. Only two of the once large family now remain. His brother, T. M. Camburn, of Tecumseh, and his sister, Mrs. S. V. Hollenbeck, of Alma, Mich. Besides these he leaves his wife and four children. Geo. M. and Edith., who live at home, Rev. Arthur T. Camburn, of Azalia and Mrs. Chas. M. Church, of 58 Dennis street. He also leaves five young grandchildren.

Mr. Camburn is a Past Commander of the local G.A.R. Post, and also a member of Blue Lodge No. 19 of the Masons.

BONUS – SHORT STORY
A CHRISTMAS TRUCE
By P.C. Zick

This short story was inspired by the *Civil War Journal of a Union Soldier.* So many of the stories my great grandfather told in his memoirs stayed with me. In uncertain times of change in the United States, his wisdom struck a chord, which I used to create this fictional story. I hope you enjoy it.

What is Christmas for a soldier such as me? I tried not to think of it. It did no good but remind us of our miserable state of affairs with the winter rains pounding down upon our heads and our huts, hastily built in the mud-covered mess of the Union army.

My family helped me along by not reminding of what I was missing, but some of the soldiers weren't as lucky as I was. They received letters from home telling them of the holiday preparations—the parties, the decorations, the baking, the gifts—all the things that would be missed sorely by those of us in the sodden misery of Virginia wearing nothing more than the scratchy wool of our winter uniforms. My mother and sisters must have known better than to send letters that would make me ache and yearn for that which could not be. At least not for Christmas of 1862, as my troop from Michigan awaited orders to march.

The winter rains had begun the week before and already roads were rutted and spirits dampened. While we waited for the rain to stop, and the war to begin again, I took little comfort in my crowded and tiny hut with its smoking fireplace, earthen floor, and cloth roof. Without comforts, conveniences, or accessories, I had nothing much to do. I knew at any time, once the rains stopped, and the sun was able to shine down on the muddy roads, all of my energies would be focused on active service.

Too much time to reflect left me wondering what it all meant. Did my family miss me, especially now that Christmas was upon them, and I wasn't there to help Father cut down the Christmas tree from my grandfather's farm on the outskirts of the small community from which I hailed? I thought back to previous years in my worst moments and remembered the party that awaited our return from the woods with the perfectly shaped tree. How could I face my rations of hard bread, bacon, and coffee when memories of sugar cookies and roasted turkey filled my senses? All the days passed one like the other in camp with our regular military duties, which amounted to very little while at rest.

After the last round of steady rain for days, we received a few supplies and a newspaper full of condemnations for the idleness of the troops in the field. But no packages from home arrived, which meant any that had been sent would not be there in time for Christmas.

Any attempt to move large bodies of men was inexpedient and to move artillery and supply trains was next to impossible with the wet and soggy conditions. The clamor of newspapers, the quarrels among general officers, and the interference of Congress with artillery movements, discouraged and demoralized our ranks. It was bad enough for some of the youngest to be away from their homes for the first time at Christmas. The men felt they were enduring hardships and sacrificing lives without adequate results and all because of petty jealousies among the leaders. Idleness and discontent go hand in hand with soldiers, and the gloomy outlook of our winter camp was not cheering. The fences had all disappeared for fuel, and green wood for cooking and heating purposes had to be hauled long distances with the mules floundering knee deep in the mire and the wagons cutting almost to the hubs.

Finally, on Christmas Eve the sun overpowered the clouds, and the incessant patter of drops on canvass stopped. I almost felt light-hearted to step outside of my hut. To break the monotony, a comrade, Jonathan, happened by and asked if I might enjoy a ride. It was the first day of sunshine we'd seen in more than a week. We both had friends in the 4th Michigan who were camped about four miles in our rear, and I decided the change of pace might very well make me miss my family less if I spent time in the company of other young men who missed home in equal measure. Our commanding officer even allowed us to take two of the horses instead of the regular mules we soldiers used for traveling with our packs. Both Jonathan and I had done extra picket duty on the stormiest nights, so we were in good stead with our superiors.

The day was filled with laughter and boasting and sunshine, and we enjoyed our visit very much. One of the soldiers told a story that had a somewhat sobering effect, although there were humorous aspects to it.

The soldier had heard about a lieutenant camped near Fredericksburg who had become enamored of a young woman who lived in an old-fashioned brick house with her mother.

The young lieutenant, whose duties called him to visit them, became acquainted with the young lady, and at her invitation called frequently upon her. He became quite taken with her charms after only a few visits that were social in nature. It wasn't usual considering both of their ages.

"Was she Confederate or Yankee?" Jonathan asked.

"It seemed he never bothered with that formality," came the storyteller's response. "He said later that because of her friendliness, he assumed her to side with us."

He continued to tell us that the lieutenant proposed marriage, and the young lady accepted with the blessing of her mother.

"Not a long courtship that," one of the soldiers said. "But then if she was charming, why wait?"

We all laughed, but when we'd settled down, the story continued.

"One evening while calling upon his intended, during a brief lull in the conversation, the heavy atmosphere bore to his ear what he judged to be the click of a telegraphic instrument," Samuel continued. "Instantly, his interest and loyalty were awakened and a suspicion of treachery aroused. Without betraying that he had heard the sound, he chatted on, his keen ear strained to catch and locate the clicking."

"How could he ever suspect his beloved?" I sang out in a high-pitched tone.

"It is wartime, gentlemen," Jonathan said. "Never trust a soul, especially an innocent maiden."

The rest shushed us and urged for the story to continue.

"At the usual hour he left, convinced that a contraband communication was going on with the enemy," Samuel said. "The next evening, taking with him a strong guard and leaving them in the yard, he again called upon the young lady."

We listened attentively to the rest of the story. Receiving him with the warmth of an expected bride, the young woman conducted him to a sofa, where clasped in each other's arms, they indulged in fond caresses and endearing words until the ominous sounds of the clicking telegraph again greeted his ear. Excusing himself for a moment that he might clear the phlegm from his throat, he opened the door and motioned vigorously to his guard despite the darkness. While the door was still open, the guard pressed in and exhibited an order from General Burnside to search the house.

"That ended the kissing, that is to be sure," one of the soldiers said. "What happened then?"

"Everything changed in an instance, it did."

The young lady, so recently the devoted lover, became a tigress. With flushed cheeks and blazing eyes, she let loose a torrent or rage and abuse upon the Union soldiers.

"Yankee brutes, Lincoln hirelings, scum of the North, and cutthroats" were hurled at the men as she let loose her hatred of the Union. Familiar with the favorite expressions of southern ladies, the guard with due deliberation proceeded with the search. Down in the cellar, they unearthed a young man with complete telegraph offices, the wires leading underground to Fredericksburg. They brought the cringing knave up into the habitable world, and he pleaded piteously for his cowardly life. The sight of his abject fear aroused the genuine affection of the young lady, and she begged in tears with the lieutenant to spare the life of her dear husband.

"A married woman!" I said. "And here she thinks we're brutes?"

It seems that she had played lover to the lieutenant for the sake of the little information she could squeeze out of him for the use of the rebels.

This was only one such story I'd heard since joining the cause very early in the war. There were many instances where southern women served as decoys, and then their men were taken prisoner. Some were even taken to their deaths. They did not hesitate at anything, if they could cripple a Yankee. As a reasonable man, I knew that the same thing might exist on the other side, if given the chance. Neither side was exempt from fighting the battle of war however they might be able to win.

Jonathan and I soon made our good-byes as we knew the light of day would soon be gone. At least we'd found a way to forget about being away from home on Christmas Eve. As we rode away, I felt pleased with my decision to leave camp for a few hours. But dark clouds descended when we were gone not much more than a mile. At first, I thought we'd stayed too late and nightfall descended upon us.

The rain began in great big dollops of water, and then came faster until we were hard pressed to see the rutted road before us. When we met a group of officers on horseback, who were shouting and obviously had enjoyed some Christmas spirits, I struggled to keep my horse steady. They shouted insults to us when we ignored them.

"Too stuck up they are," one said.

"They couldn't win this war any better than two pups still sucking on their mother's teats," hurled another.

Jonathan and I concentrated on the narrow roadway. I worried that my horse might take a wrong step and end with us both in the ditch. We passed by them without giving any mind to the officers. One of them turned his horse back toward us after we passed.

"Why did you not salute your superior officer?"

"We weren't aware that we must salute every jackass we meet," my friend said.

I secretly applauded the rejoinder, but hoped it wouldn't lead to an altercation. We hadn't meant any disrespect, but were concentrating on passing without incident with our horses since the road was rutted from the rains of the previous weeks, and there was a precipitous drop off to our right.

In great rage, the officer demanded our names with regiment and company. These we truthfully gave him. He was young and green, and probably quite drunk, or he would not have turned back for such a condescending purpose. It was bound to be a very long war indeed for someone demanding salutes in precarious or even dangerous situations. It made me wonder how we could defeat the Confederacy if we practiced warfare amongst our fellow soldiers.

"I fear the winter rains have returned," Jonathan shouted to me as he drew abreast.

"If this keeps up, it will be even more impossible to get supplies," I said. I peered through the rain that had only let up a bit and saw flickering on the other side of the field to the south of us.

"Jonathan, look over there!" I pointed to the light.

"It's a house. It may be filled with Confederates, but what have we to lose?"

"Just don't be taken in by any fair maidens." I led my horse across the field and toward the warming light of Christmas Eve candles and fires.

As we drew closer, I could see that it was a modest farmhouse, but the candles on the Christmas tree blazed from the front window. We tied up our horses to the front porch railing. A small barn stood behind the house, but I could just make out its outline in the cloud-filled gathering dark. A woman opened the front door. She walked out onto the porch, all the while peering at us.

"You're not the doctor," she said. "Who are you, and what is your business here?"

"We're about two miles from our camp," I began. "It began pouring, and our horses were having trouble on the road."

"You're Yankees." She spoke in a flat voice. We would not be welcomed here.

"We are, but we mean no harm." Jonathan pulled a white handkerchief out of his pocket and waved it above his head.

"Are you expecting a doctor?" I asked. "You seemed surprised that we weren't the doctor."

I wanted desperately to climb the steps to the covered porch, but she was not welcoming.

"My sister is in labor, and we sent for the doctor hours ago."

"The roads are terrible." I swept my arm out over the rain that had started to pick up. "How far apart are the pains?"

She pursed her lips. She didn't want to respond, but then I heard a noise from inside, and she turned her head toward the front door.

"Five minutes, maybe closer together by now."

"I spent much of my childhood on my grandfather's farm," I began. "I don't know much about humans, but I've assisted on plenty of births of our animals. I could perhaps provide some assistance."

Her face went through a gambit of emotions until worry for her sister seemed to win out.

"I suppose I don't have any choice. I've never seen anything born before in my life."

"My name is William Bradford, and this is Jonathan Cameron." I took a couple of steps toward the door, and then considered what might put her most at ease. I pulled my rifle from my shoulder and set it down on the step. Jonathan did the same thing.

"I'm Susanna Wolfson. Please come onto the porch where it's dry while I warn my sister. She'll be none too pleased that her help comes in the form of a Yankee soldier."

We waited in the cover of the porch while our clothes dripped. She soon returned with towels.

"I've asked the house maid to rustle up some dry clothes. My father recently passed, and I'm sure there is something in his room that will do for now."

After I'd changed into some dry, albeit large clothes, Susanna led me into a darkened bedroom at the top of the stairs. I found the sister, Elizabeth, in the throes of a labor pain. A Negress, I assumed a slave, stood fanning her.

"How long since the last one?" I asked her.

"Four minutes gone."

I nodded and turned to Susanna. "Do you have someone who can start the boiling of water and making us compresses?"

"We have water boiling."

I asked them to bring me hot towels that could be laid on her swollen belly.

"You're a Yankee," Elizabeth muttered from her bed once the pain stopped. "Are you going to cut my baby out of me and leave me to die?"

"Of course not," I said. Her question left me nonplussed, but I supposed not out of order, when to her mind, I was the enemy.

"There are no gentlemen in the Yankee army," Elizabeth said through clenched teeth. "You are all villains and cutthroats."

"I assure you, I was raised to respect all living things," I said. "It's this war that has caused us to be enemies on opposite sides of the field. I have no intention of anything other than helping you bring your child into the world."

"Even if I name him Johnny Reb?"

"Even if you name him Jefferson Davis."

That brought a smile to both of the sisters. Finally headway.

"From what I know of the birthing process, it will still be some time before your little Johnny makes his way into the world. I'll leave you for now. Try to rest when you can."

Susanna and I walked out into the hallway.

"You and your friend must be hungry. We have the remnants of our supper that we can share."

"That would be surely appreciated."

Jonathan and I sat at the kitchen table eating the pork and potatoes laid out before us. There was cornbread as well. It was the best meal we'd seen in weeks, and we made it disappear in no time.

"We hate the Yankees, you know." Susanna poured us steaming cups of coffee. "You may be acting like gentlemen right now, but I have no faith that you won't rob us blind before you leave."

"Have you known any Yankees before tonight?" I asked.

"No, but we've heard all the stories. Yankees have no regard for the dignity of life. You are scourges upon the earth."

I saw Jonathan squirm in his seat. I struggled to keep my temper. I even managed to smile at her pronouncement.

"So I take it all the Confederate soldiers are gentlemen?" I asked in as mild a tone as I could muster under the circumstances.

"Yes, every single one."

"Think again. Is there not at least one man in the Confederate army whom you would hesitate to associate with?"

"Well, yes, perhaps one." Susanna's response came slowly, but at least there was the opening I wanted.

"Now, really aren't there many?" I asked

She looked at me with a frown. I thought I might have stepped over a boundary, until she responded.

"Well, I'll be honest with you. There are many, but most of them are gentlemen."

"That is exactly the case with the Yankee army." I had gotten through to her. "The great majority of its numbers are gentlemen, but it is to be regretted that a few are not, and tonight maybe we'll prove the truth of this statement."

"He's right, you know," Jonathan interjected. "Just tonight we were almost run off the road by a Yankee officer who thought we should have been saluting him instead of keeping our horses from falling into a ravine. We might still be court martialed since he took down our names."

Susanna stood and began pacing. "It's so hard when all you hear are the horrible things, and we're all on edge right now."

"That's what war does," I said. "It's even harder when we're fighting our fellow countrymen. Do you know sometimes when we're out on picket on quiet nights, either one of us or one of the Confederate soldiers will raise a white handkerchief, and then we'll both come to the line to pass the night away in conversation?"

"That's hard to believe." Susanna stopped pacing and sat down at the table.

We heard commotion at the front door and went with Susanna to see what might be happening. Relief flooded through me, when she greeted the man as Dr. Johnson. I wouldn't have to birth a baby after all. She led him upstairs, but when she came back down, she invited us into the parlor. She went to the piano.

"It always calms me down to play, but I'm afraid I only know Confederate songs."

"We will take no offense, but will enjoy the entertainment," I assured her.

She played the *Confederate Wagon*, the *Bonnie Blue Flag* and others. Afterwards, she whirled upon the piano stool to face us.

"You have been so kind, I think I will play the *Star Spangled Banner* for you."

By the time she had finished, the rain had stopped. Jonathan and I decided we should head back to our camp. All appeared calm in the upper region of the house.

"Thank you, Susanna," I said as we prepared to leave after donning our damp uniforms. "It has been a pleasure to meet a true southern lady."

"And I to meet two Yankee gentlemen." She grasped my hand to shake it. "I shall tell Elizabeth to keep the faith that her husband may be in the hands of men such as you."

"What do you mean?" I asked.

"He was taken prisoner of war in Fort Lafayette last month," she said. "We've heard nothing since then."

"I shall look into this," I promised. "And send word either in person or through a courier as to his well-being."

"Then please stay the night until the baby is born so you may send him word that he has a child."

"Nothing would please us more," I said.

As we settled on the living room floor for a dry night's rest, I reflected on our day. I suddenly remembered that in a few hours it would be Christmas.

"Merry Christmas, Jonathan. It may not be home, but we've all been given a great gift tonight."

"What's that?"

"We've all learned that we are much more than this pointless war."

And as we drifted off to sleep, the strains of a baby's cries wafted down the stairs. New life pulsed as night settled over us, and I fell asleep with hope for the first time in almost two years.

THE END

I am very appreciative that you took the time to read *A Civil War Journal of a Union Soldier*, and I hope you enjoyed it as much as I enjoyed producing and publishing the words of my great grandfather, Harmon Camburn.

If you did enjoy it, please recommend the book to others and leave a review of the book on Amazon or Goodreads. Thank you.

Patricia Camburn Zick

SOURCES FOR ANNOTATIONS

Printed Material
The Civil War Top Ten by Thomas R. Flagel. New York: Bristol Park Books, 2010.
The Civil War – The Conflict That Changed America, National Geographic Special Edition, Summer 2013.

Websites
Civil War Trust – www.civilwar.org
The Civil War - www.sonofthesouth.net
The History Place – www.historyplace.com
Military Factory - www.militaryfactory.com

ABOUT P.C. ZICK

P.C. Zick describes herself as a storyteller no matter what she writes. And she writes in a variety of genres, including romance, contemporary fiction, and nonfiction. She's won various awards for her essays, columns, editorials, articles, and fiction. Currently, crafting fiction—mostly romances—occupies her time.

Many of her novels contain stories of Florida and its people and environment, which she credits as giving her a rich base for her storytelling. She says, "Florida's quirky and abundant wildlife - both human and animal - supply my fiction with tales almost too weird to be believable." The three novels in her Florida Fiction series explore rural, state, and global politics, address the fight between environmentalists and developers, and capture the lives of the people struggling to survive it all. With touches of humor and romance, these novels trace Florida's history, delve into current events, and imagine future impacts of both.

The three novels in her Behind the Love trilogy are also set in Florida. From love at first sight, to giving love a second chance to friends turning to lovers, the novels run the spectrum of the contemporary romance genre.

Ms. Zick's Smoky Mountain Romances are all set in southwestern North Carolina and filled with sweet novellas perfect for retreating into the lives of people living in the mountains.

Her latest endeavor of sweet romances, Rivals in Love, takes readers to Chicago and the Crandall family as six siblings of a famous politician attempt to find romance. The first book in the series, Love on Trial, will be released in early 2017 with Love on Board to follow in the spring.

No matter the genre, her novels contain elements of romance with strong female characters, handsome heroes, and descriptive settings. She believes in living lightly upon this earth with love, laughter, and passion, and through her fiction, she imparts this philosophy in an entertaining manner with an obvious love for her characters, plot, and themes.

Visit her website to find out more about her writing life at http://www.pczick.com.

Made in the USA
Lexington, KY
06 May 2017